True Cost Accounting

This book explores the historical and theoretical classification of True Cost Accounting (TCA) and its business-oriented ways of implementation.

Tobias Gaugler, Jennifer Kunz, and Amelie Michalke provide compact and comprehensible insights into TCA, which is one approach used to comprehensively assess actual (human or natural) resource consumption and implement this assessment into companies' internal controlling. To achieve this goal, the volume locates this approach in existing economic theory and shows its historical roots and progressions. The authors then present different methodological approaches within TCA and highlight how practitioners might implement these approaches in their business. To illustrate the importance of TCA, a number of practical examples that are already present in the market are also featured.

This volume will be of great interest to students and scholars of environmental economics. Furthermore, practitioners in (sustainable) supply chain management, as well as ecological and social accounting and controlling, can use it for inspiration and guidance. Additionally, this book and the inherent ideas will give policymakers in economic and social politics recommendations for action.

Amelie Michalke is a postdoc at the Nuremberg Institute of Technology, Germany, and the co-founder of Patos, a consultancy for life cycle assessment and TCA.

Jennifer Kunz is a full professor and the head of the Chair of Management Accounting and Management Control at the University of Augsburg, Germany.

Tobias Gaugler is research professor at the Nuremberg Institute of Technology, Germany, and head of the Competence Center for Sustainable Food and Resource Management.

Routledge Focus on Environment and Sustainability

Sustainable Marketing and the Circular Economy in Poland
Key Concepts and Strategies
Anita Proszowska, Ewa Prymon-Ryś, Anna Kondak, Aleksandra Wilk and Anna Dubel

Risk Management for Water Professionals
Technical, Psychological and Sociological Underpinnings
Anna Kosovac

Climate Perspectives from the Congo Basin
Bila-Isia Inogwabini

Everyday Agri-Environmental Governance
The Emergence of Sustainability through Assemblage Thinking
Jérémie Forney, Dana Bentia, and Angga Dwiartama

Electric Cars and the Resource Challenge
Theo Henckens

Regulating Foreign Direct Investment for Development
Perspectives from Bangladesh
Nakib Mohammad Nasrullah and Mia Mahmudur Rahim

True Cost Accounting
Assessing Environmental and Social Damages of Production
Amelie Michalke, Jennifer Kunz, and Tobias Gaugler

For more information about this series, please visit: www.routledge.com/Routledge-Focus-on-Environment-and-Sustainability/book-series/RFES

True Cost Accounting

Assessing Environmental and Social Damages of Production

Amelie Michalke, Jennifer Kunz, and Tobias Gaugler

Routledge
Taylor & Francis Group
LONDON AND NEW YORK

earthscan
from Routledge

First published 2025
by Routledge
4 Park Square, Milton Park, Abingdon, Oxon OX14 4RN

and by Routledge
605 Third Avenue, New York, NY 10158

Routledge is an imprint of the Taylor & Francis Group, an informa business

British Library Cataloguing-in-Publication Data
A catalogue record for this book is available from the British Library

ISBN: 978-1-032-38466-5 (hbk)
ISBN: 978-1-032-38467-2 (pbk)
ISBN: 978-1-003-34519-0 (ebk)

DOI: 10.4324/9781003345190

Typeset in Times New Roman
by KnowledgeWorks Global Ltd.

Contents

Acknowledgments

The authors would like to extend their heartfelt thanks to Benjamin Oebel and Lennart Stein for their extensive expertise and support, without which this book would not have been possible. Their gratitude also goes to Tim Andreae, Paul Glöckner, and Maximilian Pieper for their valuable contributions.

The authors are especially grateful to Rosie Anderson, Jyotsna Gurung, and Serena Watkins from Taylor and Francis for their smooth, thorough, and patient collaboration.

Part I

Overview

1 Bridging the Gap between Ecology, Society, and Economy

Abstract

Environmental destruction and social injustice induced by the mechanisms underlying the global economic system constitute an existential threat to the world's ecosphere and its people. An ever-growing global population further increases this claim on resources, especially in countries of higher wealth and power. What is needed on a grand scale are measures, which combine all three pillars of sustainability: Ecology, society, and economy. This can be achieved if actual resource consumption and its following damages are comprehensively assessed in ways that companies can understand and implement into their internal controlling and if doing so does not remain a voluntary corporate decision. True Cost Accounting (TCA) is an assessment approach that aims to leverage societal and ecological – and, in turn, economic – benefits from allegedly adjusted demand patterns due to changed pricing of commodities. The present chapter positions TCA in relation to the current social discourse about sustainability and resilience and assigns its importance to contribute to solving global issues.

Key takeaways

After reading this chapter, you will:

- Consider environmental and social issues as consequences of current unsustainable economics
- Understand how these social and ecological issues have economic implications in turn
- Learn about global goals and approaches for sustainable development
- Be introduced to the concept of True Cost Accounting as a tool for holistic business reporting and sustainable transition of consumption and production

1.1 Sustainability in the Anthropocene

Key issue: How is the economy subject to environmental and societal sustainability?

DOI: 10.4324/9781003345190-2

Environmental destruction and social injustice induced by the mechanisms underlying the global economy constitute an existential threat to the world's ecosphere and hence its people. Being part of a global society was painfully obvious during the COVID-19 pandemic, which presented as possibly the first – at least in living history – issue that penetrated seemingly every corner of the globe and let us face one communal threat: Our jeopardized health. But the communal threat of the planet's jeopardized health is and was omnipresent, even if we don't see it as such. We seem disconnected with the idea that the individual problems we may face – extreme weather events, poverty and world hunger, health crises – are targeting us as one common people (Spence et al. 2011, Olmos 2020) and are all attributable to one common source: Human action within a globally unsustainable economic system (Wijkman and Rockström 2013, Wiedmann et al. 2020).

Not least, the *Intergovernmental Panel on Climate Change* (IPCC) stipulates the importance of drastically limiting greenhouse gas (GHG) emissions from anthropogenic sources, or else face major environmental destruction yet unknown to mankind (IPCC 2022). A rising number of extreme weather events are observable in Europe (Kron et al. 2019). To a great extent, these are attributable to anthropogenic GHG emissions (Otto et al. 2020), a fact that is important to communicate for public awareness of the urgency to reduce man-made emissions (Schiermeier 2018). And public awareness seems to be on the rise: Also due to the efforts of fierce youth taking to the streets around the globe as the movement *Fridays For Future* to protest a lack of political action against the climate crisis, climatic breakdown is at the forefront of public discourse on environmental health (Beckh and Limmer 2022). Climate issues affect human well-being not only from an environmental perspective but also economically: For example, the insurance industry, as the financial risk manager of our society, must take the likelihood and severity of climate change effects into account and adapt capital provision or investment accordingly, which could likewise affect the insured (Hecht 2008, Herweijer et al. 2009, Jarzabkowski et al. 2019). However, other pressing matters further endanger human societies of today and tomorrow: Rockström et al. (2009) introduce the concept of planetary boundaries, which describe environmental limits that define the "safe operating space" of contemporary societies. Besides climate change, these include, e.g., the loss of biodiversity, the nitrogen and phosphorus cycles, ocean acidification, or land system change. A total of nine boundaries have been updated in Steffen et al. (2015) with an observable transgression of environmental systems toward high-risk zones, especially in nutrient cycling of phosphorus and nitrogen, and the planet's genetic diversity. O'Neill et al. (2018) further find that no country worldwide can meet their population's needs without overshooting levels of resource use and therefore neglecting the planetary boundaries. Authors suggest that universal human needs could safely be met, but only if sustainability goals focus on "sustainable and equitable human well-being" rather than material consumption and associated economic growth. Because what is often failed to be recognized is how environmental threats are inherently linked with economic deterioration as well. For example, improving the use of nitrogen for fertilization would result in drastic cost savings for the treatment of nitrogen-polluted waterways or savings in fertilization expenses (Langholtz et al. 2021). Furthermore, experts of the *World Economic Forum* identify five out of the ten most severe global risks over the next ten years as being of environmental origin. Especially climate action failure, extreme weather, and biodiversity loss are postulated as the "gravest threats to humanity", and tackling them is seen as more and more challenging, especially since societal issues seem to have progressed faster since the pandemic (World Economic Forum 2022).

As projections of climate and environmental development show, without drastic policy action, it is only getting worse from here on for all people, but especially those in developing countries (OECD 2012, IPCC 2022). Increased environmental pressure is also sometimes argued with an ever-increasing global population that pushes the claim on natural resources. Evidence exists for both affluent (Weber and Sciubba 2018, Khan et al. 2021) and aspiring regions (Maja and Ayano 2021). This has a detrimental impact toward environmental stability, as, for example, demand for resource-intensive food increases agricultural land and other resources required for its production (Tilman et al. 2011, Kastner et al. 2012). However, it is too easy to blame environmental destruction on population growth alone. In countries with falling or stagnant population growth – upper-middle-income and high-income countries – GHG emissions have more than doubled in the past two decades owing to intensive consumption patterns (UN 2019). While low-income countries strive to close this gap, another driving force therefore is pursued economic growth, although scholars argue how an increase in GDP through resource depletion does not necessarily increase per capita wealth (Kurniawan and Managi 2018, Adejumo 2019). Moreover, an unequal playing field with better access to resources and less threat of environmental breakdown for more privileged parts of the global population fosters environmental injustice (Agyeman et al. 2016): It first exerts environmental pressure on underprivileged strata of the population, for example, with land use change or cropland expansion in developing nations (Molotoks et al. 2018, Hentschl et al. 2023), the exploitation of countries' natural resources for minimal profit for its population (International Resource Panel 2019), or the export of toxins to the developing world (Clapp 2010); second, the damages of environmental destruction are felt to a more gravely in developing countries. Bathiany et al. (2018) find that countries, which contribute rather little to the global climate crisis, are prone to experience climate variability to a greater extent than high-emitting countries. This does not only affect the food security of these countries (McMichael et al. 2007, Karki and Gurung 2012, Thornton et al. 2014, Anderson et al. 2020, Davis et al. 2022) and their export rates, especially of agricultural goods (Jones and Olken 2010). It also hinders the countries' economic growth rates (Dell et al. 2012) further reinforcing existing economic power imbalances (Anderson et al. 2020) and hence obstructing the establishment of global environmental and social justice.

To counteract the developments of climate breakdown, ecosphere degradation, and following societal divide, some international efforts strive toward a more sustainable state of the world. To not transgress climate warming of 1.5°C at most compared to preindustrial levels, we need to cut GHG emissions almost in half by 2030 (IPCC 2022). This is only feasible by drastically increasing climate policy efforts for a fossil-free future with circular economies and ultimately a livable planet. In all, 196 parties committed to this goal of limiting global warming to well below 2°C by signing the *Paris Agreement* in 2015. With mandatory contributions of nationally determined actions (NDCs) to reduce GHGs and to build resilience toward rising temperatures, and non-mandatory long-term GHG reduction strategies, the Paris Agreement aims at holding countries accountable to pursue this common goal. That current implemented efforts are insufficient, however, shows the actual lethargy – or is it indignation? – of policies tackling climate change (Rogelj et al. 2016, Meinshausen et al. 2022, Matthews & Wynes 2022). If the *Glasgow Climate Pact* from 2021, with its updated and strengthened NDCs and financial bolstering of climate efforts, will have a more positive effect on climate stabilization remains to be seen at the time of writing. Further commitments and declarations have been put forward. Both in the United States and the European Union, a *Green New Deal* has been introduced as a legislative proposal for the transition to a sustainable, equitable, and just economy. The proposals address issues of, i.e., resilience against climate change, improving infrastructure, or job creation in renewable energies in the US proposal, and foremost the creation of a net-zero emissions economy in the EU proposal (Scheulke-Leech 2021). The EU Green New Deal is seen as pivotal to reach the *Sustainable Development Goals* (SDGs) stipulated by the United Nations (UN) and first introduced in 2016 (UN 2016). These 17 objectives include goals for sustainability in natural (e.g., 13: Climate action), economic (e.g., 8: Decent work and economic growth), and societal (e.g., 1: No poverty) systems and have been adopted in 2015 by all UN member states with a pledge to achievement until 2030. The goals were defined in cooperation and discourse between governments, companies, societal organizations, and academia, as well as citizen engagement to comprehensively describe the global stipulation of sustainability (Tulder et al. 2021). They are divided into a total of 169 targets, and their degree of achievement is measured with 231 different indicators. One example would be:

SDG 1: No poverty.

Target 1.1: By 2030, eradicate extreme poverty for all people everywhere, currently measured as people living on less than $1.25 a day.

Indicator 1.1.1: Proportion of the population living below the international poverty line by sex, age, employment status, and geographic location (urban/rural).

As the UN has acknowledged the speed toward achievement of SDGs in 2030 as too slow, they declared a *Decade of Action* starting 2020 (UN 2019). Especially human-development targets are projected to fail achievement with especially countries of Sub-Saharan Africa at risk of insufficient development (Moyer and Hedden 2020). One hindering factor is potentially the trade-offs that some goals have with one another (Nilsson et al. 2016, Fonseca et al. 2020). For example, intensifying agricultural production to reach Goal 1, zero hunger, would put more strain on natural resources and hence negatively influence Goal 6, clean water and sanitation, or Goal 13, climate change. Adequately addressing these interdependencies within the complex goals for sustainability requires great levels of interdisciplinary and systems thinking (Reynolds et al. 2017). Nilsson et al. (2016) argue that the mere design of policy making prevents an adequate evaluation of the interactions of the goals and a sensible implementation of strategies to achieving them since policies on different systems are designed and decided by different entities. This emphasizes the extent to which global systems would need to change and on how many levels change must be implemented to achieve holistic sustainability – provided that the achievement of the SDGs describes holistic sustainable development.

Businesses are recognized as key players in the achievement of the SDGs (Folke et al. 2019, Mio et al. 2020). PwC (2019) finds that 72% of over 1000 surveyed companies in 31 countries across seven industries mention SDGs in their reporting publications. But a mere mention does not equal implemented measures toward the goals' achievement. There are, however, practice-oriented sustainability approaches that indeed foster an implementation of measures toward a company's sustainable development. The EU parliament and its member states have, for example, adopted the *Corporate Social Responsibility Directive* in 2014 for more transparency about ecological and social aspects of big corporations (BMAS 2022). Just in time for the start of the Decade of Action, some laws become mandatory from 2020 onwards. For example, prospectively from 2023, corporations, banks, and insurance companies will have to follow the *Sustainability Reporting Standards*, a new accounting directive of the European sustainability policy. It obliges these actors to report nonfinancial data based on double materiality: Sustainability aspects should hence be reported from both outside-in and inside-out perspectives. The former describes which effects environmental or social aspects have on corporate success; the latter describes how corporate action affects the environment and society (Behncke 2022). These revised and improved obligations for business management are likely to affect economic design but further action has to be taken to reach an economy that truly benefits the planet and all its citizens in the long run. Because if not the satisfaction of human needs and wants globally, what is the purpose of economic activity?

Some alternative economic models have been proposed, acknowledging the intertwined ecological and social perspectives. The Council of the EU, for example, recognizes the *Economy of Wellbeing* as "a policy orientation and a governance approach, which aims to put people and their wellbeing at the centre of policy- and decision-making". This necessitates investment in public services like health services, social protection, or education (Council of the EU 2019). The *Organisation for Economic Cooperation and Development* (OECD) defines the Economy of Wellbeing as a "virtuous cycle" where the well-being of the individual positively affects the success of an economy and vice versa with the help of purposeful policies (OECD 2019). One step toward this is the measurement of well-being to find areas of life in an economy worth improving. One such measurement has been proposed by the OECD with the *Better Life Index*, allowing for the mapping of an individual's well-being across several socioeconomic aspects and the aggregation toward countries, for example (Durand 2014). Countries also engage in alternative economic design: Bhutan was the first country to introduce another measure, the *Gross National Happiness Index* (GNH), as an alternative to the GDP measuring the success of an economy (Sithey et al. 2015). While the Economy of Wellbeing focuses foremost on social dimensions, the GNH acknowledges ecological sustainability as vital for people's well-being too. Similarly, the *Doughnut Economics*, introduced by Kate Raworth as more of a new school of thought, balances the "safe and just space for humanity" and therefore economic activity, between a social foundation of critical human deprivation and an ecological ceiling of critical planetary degradation (Raworth 2017). It sets out to reframe conventional economic thinking by drawing on insights of complexity, ecological, feminist, institutional, and behavioral economics. The *Economy for the Common Good*, another economic model first described in 2010, measures an economy by its inherent ecological responsibility and social justice (Felber and Pettifor 2019). It sets out to renovate the understanding of economy in education, changing business practice in the economy, and motivating legislative reform in economic politics (Felber XY). With the *Common Good Matrix* and *Common Good Balance Sheets*, businesses participate in the Economy for the Common Good working as a sustainability management and control system (Ejarque and Campos 2020). About 400 businesses currently operate under the principles of this economic model currently (Felber and Pettifor 2019). Considering the challenges of modern life on our planet, a revolution in economics seems urgently needed and likewise underway with engagement in the presented and further models, like *post-growth economics* (Latouche 2009) or the *caring economics* (Eisler 2008).

1.2 An approach to bridge the gap: True Cost Accounting

Key issue: How can TCA foster sustainable development?

Alternative economic models are walking away from a focus on endless GDP growth. Instead, economic performance is measured in accordance with environmental and societal benefits. These models are helpful to conceptualize the conditions in which a sustainable form of economy is to develop. But beyond models, what is needed on a grand scale are implemented measures, which integrate ecological and societal quests into economic practice.

Efficient and fair levels of resource use – be that natural or human resources – have to be reached to increase societal (climate)resilience and overall sustainability. Since economic action is settled on "the market", one way of striving for this efficiency and equity is a comprehensive assessment of actual resource consumption and its following damages in the language of the market: Money. Measures that draw on the economic language of sustainability are, for example, indicated under SDG 17 "Partnership for the goals". Target 17.19 describes that "measurements of progress on sustainable development that complement GDP [...]" should be developed. Bebbington et al. (2020) recognize that in an era shaped by anthropogenic impacts – or the era of the Anthropocene, if you will – business organization and management will have to adapt accordingly. This might imply that managerial and operational tools, like tools of accounting, also must develop along with the complexity of global environmental change and its sources. To date, these anthropogenic impacts are investigated exclusively by how they affect an economic system. This makes business as usual highly unsustainable, certainly from an environmental and social perspective, but also since it evokes unaccounted economic impact from damages beyond financial consideration. It also risks the future business success of rigid companies that might not be able to keep up with sustainability endeavors in ever-evolving markets (Hopwood et al. 2010). With an integration of environmental and social interactions, however, a more holistic picture of the responsibilities of anthropogenic activity is drawn. Such currently unaccounted interactions are also described as "externalities", since they cause costs that are externalized and eventually borne by actors uninvolved in their source's financial transactions (Hopwood et al. 2010, Unerman et al. 2018).

True Cost Accounting (TCA) is one measure that aims at balancing all negative and positive impacts accumulated along the value chain of a product, or the activity of an enterprise or company, and with it especially the impact that is not accounted for in current pricing mechanisms (TEEB 2018, Baker et al. 2020, Castilleja 2021, True Cost Initiative 2022). With TCA, externalities are valued monetarily as "hidden costs" or "hidden benefits" and hence visualize market distortions currently present (Baker et al. 2020) or determine the level of sustainability of an enterprise (True Cost Initiative 2022). Another term of accounting for environmental and societal costs has emerged in literature over the past decades. Full cost accounting (FCA) – or sometimes full environmental cost accounting – was introduced by the European Commission in 1992 and further explored by Bebbington et al. (2001). It is described as a tool to correct prices by including transactional costs and benefits that are hitherto ignored in business decisions – or for short, externalities. Authors, however, stipulate that the term "full cost" comes with caveats since it is presumptuous to assume all complexities of environmental or societal issues can indeed be measured to the cent. This is a notion that prevails in TCA literature until today (Michalke et al. 2022). FCA has been applied across different industries, like coal (Epstein et al. 2011), waste (D'Onza et al. 2016), or timber (Herbohn 2005). To find how FCA can be applied in the automotive case, Jasinski et al. (2015) conduct a literature review of FCA and the used full cost calculation methods. One such method is the monetary valuation of life cycle assessment (LCA) results. This is the route that has been taken recently in TCA literature, which has foremost been applied in research of the agricultural or food sector (True Cost Initiative 2022, Michalke et al. 2022) but can, of course, be used regardless of the industry a product or company belongs to. The use of "true costs" instead of "full costs" might come across a little more complacent in conveying how these costs are hidden in current pricing scenarios. However, backgrounds, methods, and the use of FCA or TCA results are equivalent. As the title of this book suggests and since recently in literature this term has been predominantly used, we talk about "TCA" in this book.

TCA is used for different purposes, like the consulting of companies for their environmental accountability (KPMG 2012), transparently communicating business activities and finding potential to minimize impact (EOSTA 2017, True Cost Initiative 2022), or engaging with the general public through informational campaigning (Michalke et al. 2020). In this book, we will approach TCA foremost from the perspective of business reporting. We want to draw a case of TCA that can be understood and implemented in companies' internal controlling. However, while TCA is one path toward new standards of business reporting, we also emphasize how it affects consumers' decision processes and outcomes, and therefore the potential it holds of enhancing sustainable consumption and production patterns. For customers with financial decisions strongly influencing purchasing behavior (Seubelt et al. 2022), consumption patterns possibly shift toward more sustainable goods since after an implementation of TCA measures or their true costs calculated these products are also more favorable from a financial standpoint. Chapter 2 will explore the economic theoretical background and how demand changes with price fluctuations. It will also touch upon policy measures for implementation. The methodological approach of calculating "true" costs, and therefore the act of monetizing environmental and societal impact, will be explained in Chapter 3. In this book, we will approach the method from the perspective of an emerging strand of research that is particularly concerned with TCA in the food sector (e.g., Sustainable Food Trust 2019, Pieper et al. 2020, Michalke et al. 2022). However, we will only use the methods of TCA developed therein and present them in a way that can be conceptualized for all other fields of producing industry likewise. With mandatory accounting for – and eventually compensation of – externalities, companies are incentivized to minimize their environmental and social burden from two perspectives: Their stance in society and therefore with their customers, and their economic interest in minimizing overall costs. Therefore, Chapter 4 will focus on the use of TCA as part of business reporting alongside conventional cost accounting in company management or operation. While TCA has not been broadly implemented in economic policies, we want to motivate the use of it as a tool for sustainability engagement by showcasing some cases of TCA implementation in Chapter 5. Finally, we will draw concluding remarks for all stakeholders concerned and influenced by the use of TCA in Chapter 6.

1.3 Review questions

1 Find an example where ecological or social issues have direct economic impact (example from the text: Rising costs of water purification due to nitrogen pollution).
2 What are key hinderances of global environmental justice?
3 Discuss trade-offs between different SDGs.
4 What is meant with the "revolution of economics"?
5 How can TCA help foster this revolution?
6 Describe the core aim of TCA.
7 What are perspectives from which TCA can be fruitful for sustainable development?

References

Adejumo, O. O. (2019). Environmental quality vs economic growth in a developing economy: Complements or conflicts. *Environmental Science and Pollution Research, 27*(6), 6163–6179. https://doi.org/10.1007/s11356-019-07101-x

Agyeman, J., Schlosberg, D., Craven, L., & Matthews, C. (2016). Trends and directions in environmental justice: From inequity to everyday life, community, and just sustainabilities. *Annual Review of Environment and Resources, 41*(1), 321–340. https://doi.org/10.1146/annurev-environ-110615-090052

Anderson, R., Bayer, P. E., & Edwards, D. (2020). Climate change and the need for agricultural adaptation. *Current Opinion in Plant Biology, 56*, 197–202. https://doi.org/10.1016/j.pbi.2019.12.006

Baker, L., Castilleja, G., Ruiz, A. D. G., & Jones, A. (2020). Prospects for the True Cost Accounting of food systems. *Nature Food, 1*(12), 765–767. https://doi.org/10.1038/s43016-020-00193-6

Bathiany, S., Dakos, V., Scheffer, M., & Lenton, T. M. (2018). Climate models predict increasing temperature variability in poor countries. *Science Advances, 4*(5). https://doi.org/10.1126/sciadv.aar5809

Bebbington, J. (2001). Sustainable development: A review of the international development, business and accounting literature. *Accounting Forum, 25*(2), 128–157.

Bebbington, J., Österblom, H., Crona, B., Jouffray, J. B., Larrinaga, C., Russell, S., & Scholtens, B. (2020). Accounting and accountability in the Anthropocene. *Accounting, Auditing & Accountability Journal, 33*(1), 152–177. https://doi.org/10.1108/AAAJ-11-2018-3745

Beckh, P., & Limmer, A. (2022). The Fridays for future phenomenon. *Strategies for sustainability of the Earth system*, 427–432. https://doi.org/10.1007/978-3-030-74458-8_28

Behncke, N. (2022). *Doppelte Materialität – Mehr als eine regulatorische Pflichtaufgabe*. Available at: https://blogs.pwc.de/de/sustaining-values/article/232878/doppelte-materialitaet-mehr-als-eine-regulatorische-pflichtaufgabe/#:~:text=Doppelte

BMAS. (2022). *CSR Politik- Aktivitäten der Bundesregierung*. Available at: https://www.csr-in-deutschland.de/DE/CSR-Allgemein/CSR-Politik/CSR-in-Deutschland/Aktivitaeten-der-Bundesregierung/aktivitaeten-der-bundesregierung.html#:~:text=CSR-Berichtspflichten&text=Ziel

Castilleja, G. (2021). *True Cost Accounting for Food. Balancing the Scale*. (P. A. D. Barbara Gemmill-Herren Lauren Baker, Hrsg.; S. xxxi–xxxv). Routledge (Routledge studies in food, society and the environment).

Clapp, J. (2010). *Toxic Exports*. Cornell University Press. https://doi.org/10.7591/9781501735936

Council of the European Union. (2019). *Council conclusions on the Economy of Wellbeing*. Available at: https://data.consilium.europa.eu/doc/document/ST-11164-2019-INIT/en/pdf

Davis, B., Lipper, L., & Winters, P. (2022). Do not transform food systems on the backs of the rural poor. *Food Security*. https://doi.org/10.1007/s12571-021-01214-3

Dell, M., Jones, B. F., & Olken, B. A. (2012). Temperature shocks and economic growth: Evidence from the last half century. *American Economic Journal: Macroeconomics*, *4*(3), 66–95. https://doi.org/10.1257/mac.4.3.66

D'Onza, G., Greco, G., & Allegrini, M. (2016). Full cost accounting in the analysis of separated waste collection efficiency: A methodological proposal. *Journal of Environmental Management*, *167*, 59–65. https://doi.org/10.1016/j.jenvman.2015.09.002

Durand, M. (2014). The OECD better life initiative: How's life? The measurement of well-being. *Review of Income and Wealth*, *61*(1), 4–17. https://doi.org/10.1111/roiw.12156

Eisler, R. (2008). *The Real Wealth of Nations: Creating a Caring Economics*. Berrett-Koehler Publishers.

Ejarque, A. T., & Campos, V. (2020). Assessing the Economy for the Common Good measurement theory ability to integrate the SDGs into MSMEs. *Sustainability*, *12*(24), 10305. https://doi.org/10.3390/su122410305

EOSTA. (2017). *True cost accounting-pilotstudie legt wahre Kosten der Lebensmittelproduktion für Mensch und Umwelt offen*. Available at: https://www.eosta.com/de/aktuell/true-cost-accounting-pilotstudie-legt-wahre-kosten-der-lebensmittelproduktion-fuer-mensch

Epstein, P. R., Buonocore, J. J., Eckerle, K., Hendryx, M., Stout III, B. M., Heinberg, R., Clapp, R. W., May, B., Reinhart, N. L., Ahern, M. M., Doshi, S. K., & Glustrom, L. (2011). Full cost accounting for the life cycle of coal. *Annals of the New York Academy of Sciences*, *1219*(1), 73–98. https://doi.org/10.1111/j.1749-6632.2010.05890.x

Felber, C., & Pettifor, A. (2019). *Change Everything: Creating an Economy for the Common Good*. Zed Books Ltd.

**Folke, C., Österblom, H., Jouffray, J.-B., Lambin, E. F., Adger, W. N., Scheffer, M., Crona, B. I., Nyström, M., Levin, S. A., Carpenter, S. R., Anderies, J. M., Chapin, S., Crépin, A.-S., Dauriach, A., Galaz, V., Gordon, L. J., Kautsky, N.,

Walker, B. H., Watson, J. R., … Zeeuw, A. de. (2019). Transnational corporations and the challenge of biosphere stewardship. *Nature Ecology & Evolution*, *3*(10), 1396–1403. https://doi.org/10.1038/s41559-019-0978-z

Fonseca, L. M., Domingues, J. P., & Dima, A. M. (2020). Mapping the sustainable development goals relationships. *Sustainability*, *12*(8), 3359. https://doi.org/10.3390/su12083359

Hecht, S. B. (2008). Climate change and the transformation of risk: Insurance matters. *UCLA Law Review*, *55*, 1559–1620.

Hentschl, M., Michalke, A., Pieper, M., Gaugler, T., & Stoll-Kleemann, S. (2023). Dietary change and land use change: Assessing preventable climate and biodiversity damage due to meat consumption in Germany. *Sustainability Science*, 1–17. https://doi.org/10.1007/s11625-023-01326-z

Herbohn, K. (2005). A full cost environmental accounting experiment. *Accounting, Organizations and Society*, *30*(6), 519–536. https://doi.org/10.1016/j.aos.2005.01.001

Herweijer, C., Ranger, N. & Ward, R. (2009). Adaptation to climate change: Threats and opportunities for the insurance industry. *The Geneva Papers on Risk and Insurance. Issues and Practice*, *34*, 360–380. https://doi.org/10.1057/gpp.2009.13

Hopwood, A. G., & Unerman, J. (Eds.). (2010). *Accounting for Sustainability: Practical Insights*. Earthscan.

International Resource Panel. (2019). *Global resources Outlook 2019: Natural resources for the future we want*. Available at: https://www.resourcepanel.org/reports/global-resources-outlook-2019

IPCC. (2022). *The evidence is clear: The time for action is now. We can halve emissions by 2030*. Available at: https://www.ipcc.ch/2022/04/04/ipcc-ar6-wgiii-pressrelease/

Jarzabkowski, P., Chalkias, K., Clarke, D., Iyahen, E., Stadtmueller, D., & Zwick, A. (2019). *Insurance for climate adaptation: Opportunities and limitations*. Available at: https://eprints.bbk.ac.uk/id/eprint/28797/1/insurance-for-climate-adaptation-opportunities-limitations.pdf

Jasinski, D., Meredith, J., & Kirwan, K. (2015). A comprehensive review of full cost accounting methods and their applicability to the automotive industry. *Journal of Cleaner Production*, *108*, 1123–1139. https://doi.org/10.1016/j.jclepro.2015.06.040

Jones, B. F., & Olken, B. A. (2010). Climate shocks and exports. *American Economic Review*, *100*(2), 454–459. https://doi.org/10.1257/aer.100.2.454

Karki, R., & Gurung, A. (2012). An overview of climate change and its impact on agriculture: A review from least developing country, Nepal. *International Journal of Ecosystem*, *2*(2), 19–24. https://doi.org/10.5923/j.ije.20120202.03

Kastner, T., Rivas, M. J. I., Koch, W., & Nonhebel, S. (2012). Global changes in diets and the consequences for land requirements for food. *Proceedings of the National Academy of Sciences*, *109*(18), 6868–6872. https://doi.org/10.1073/pnas.1117054109

Khan, I., Hou, F., & Le, H. P. (2021). The impact of natural resources, energy consumption, and population growth on environmental quality: Fresh evidence from the United States of America. *Science of The Total Environment*, *754*, 142222. https://doi.org/10.1016/j.scitotenv.2020.142222

KPMG. (2012). *Sustainable insight. Expect the unexpected: Building business value in a changing world.* March 2012. Available at: https://assets.kpmg.com/content/dam/kpmg/pdf/2012/03/sustainable-insights-march-2012.pdf

Kron, W., Löw, P., & Kundzewicz, Z. W. (2019). Changes in risk of extreme weather events in EUROPE. *Environmental Science & Policy*, *100*, 74–83. https://doi.org/10.1016/j.envsci.2019.06.007

Kurniawan, R., & Managi, S. (2018). Economic growth and sustainable development in Indonesia: An assessment. *Bulletin of Indonesian Economic Studies*, *54*(3), 339–361. https://doi.org/10.1080/00074918.2018.1450962

Langholtz, M., Davison, B. H., Jager, H. I., Eaton, L., Baskaran, L. M., Davis, M., & Brandt, C. C. (2021). Increased nitrogen use efficiency in crop production can provide economic and environmental benefits. *Science of The Total Environment*, *758*, 143602. https://doi.org/10.1016/j.scitotenv.2020.143602

Latouche, S. (2009). *Farewell to Growth*. Polity.

Matthews, H. D., & Wynes, S. (2022). Current global efforts are insufficient to limit warming to 1.5 C. *Science*, *376*(6600), 1404–1409. https://doi.org/10.1126/science.abo3378

Maja, M. M., & Ayano, S. F. (2021). The impact of population growth on natural resources and farmers' capacity to adapt to climate change in low-income countries. *Earth Systems and Environment*, *5*(2), 271–283. https://doi.org/10.1007/s41748-021-00209-6

McMichael, A. J., Powles, J. W., Butler, C. D., & Uauy, R. (2007). Food, livestock production, energy, climate change, and health. *The Lancet*, *370*(9594), 1253–1263. https://doi.org/10.1016/s0140-6736(07)61256-2

Meinshausen, M., Lewis, J., McGlade, C., Gütschow, J., Nicholls, Z., Burdon, R., … & Hackmann, B. (2022). Realization of Paris Agreement pledges may limit warming just below 2 C. *Nature*, *604*(7905), 304–309. https://doi.org/10.1038/s41586-022-04553-z

Michalke, A., Stein, L., Fichtner, R., Gaugler, T., & Stoll-Kleemann, S. (2022). True Cost Accounting in agri-food networks: A German case study on informational campaigning and responsible implementation. *Sustainability Science*, *17*(6), 2269–2285. https://doi.org/10.1007/s11625-022-01105-2

Michalke, A., Stoll-Kleemann, S., & Gaugler, T. (2020). *Abschlussbericht zum Forschungsprojekt How much is the dish? - True Cost Accounting von Umweltfolgekosten und "wahre Preisschilder" in Deutschland. HoMaBiLe (How much is the dish? – Measures for increasing biodiversity through True Cost Accounting for food products).* Available at: https://www.researchgate.net/publication/346097024_True_Prices

Mio, C., Panfilo, S., & Blundo, B. (2020). Sustainable development goals and the strategic role of business: A systematic literature review. *Business Strategy and the Environment*, *29*(8), 3220–3245. https://doi.org/10.1002/bse.2568

Molotoks, A., Stehfest, E., Doelman, J., Albanito, F., Fitton, N., Dawson, T. P., & Smith, P. (2018). Global projections of future cropland expansion to 2050 and direct impacts on biodiversity and carbon storage. *Global Change Biology*, *24*(12), 5895–5908. https://doi.org/10.1111/gcb.14459

Moyer, J. D., & Hedden, S. (2020). Are we on the right path to achieve the sustainable development goals? *World Development*, *127*, 104749. https://doi.org/10.1016/j.worlddev.2019.104749

Nilsson, M., Griggs, D., & Visbeck, M. (2016). Policy: Map the interactions between sustainable development goals. *Nature*, *534*(7607), 320–322. https://doi.org/10.1038/534320a

OCED. (2012). *OECD environmental Outlook to 2050: The consequences of inaction—Key facts and figure*. Available at: https://www.oecd.org/content/dam/oecd/en/publications/reports/2012/03/oecd-environmental-outlook-to-2050_g1g14e69/9789264122246-en.pdf

Olmos, I. L. (2020). The space between us: Social geography and politics. *The AAG Review of Books*, *8*(4), 208–210. https://doi.org/10.1080/2325548x.2020.1781451

O'Neill, D. W., Fanning, A. L., Lamb, W. F., & Steinberger, J. K. (2018). A good life for all within planetary boundaries. *Nature Sustainability*, *1*(2), 88–95. https://doi.org/10.1038/s41893-018-0021-4

Organisation for Economic Co-Operation and Development (OECD). (2019). *The economy of well-being*. Available at: https://doi.org/10.1787/498e9bc7-en

Otto, F. E. L., Harrington, L., Schmitt, K., Philip, S., Kew, S., Oldenborgh, G. J., Singh, R., Kimutai, J., & Wolski, P. (2020). Challenges to understanding extreme weather changes in lower income countries. *Bulletin of the American Meteorological Society*, *101*(10), E1851–E1860. https://doi.org/10.1175/BAMS-D-19-0317.1

Pieper, M., Michalke, A., & Gaugler, T. (2020). Calculation of external climate costs for food highlights inadequate pricing of animal products. *Nature Communications*, *11*(1). https://doi.org/10.1038/s41467-020-19474-6

PwC. (2019). *Creating a strategy for a better world*. Available at: https://www.pwc.com/gx/en/sustainability/SDG/sdg-2019.pdf

Raworth, K. (2017). *Doughnut Economics: Seven Ways to Think Like a 21st-Century Economist*. Chelsea Green Publishing.

Reynolds, M., Blackmore, C., Ison, R., Shah, R., & Wedlock, E. (2017). The Role of Systems Thinking in the Practice of Implementing Sustainable Development Goals. In *World Sustainability Series* (S. 677–698). Springer International Publishing. https://doi.org/10.1007/978-3-319-63007-6_42

Rockström, J., Steffen, W., Noone, K., Persson, Å., Chapin III, F. S., Lambin, E., … & Foley, J. (2009). Planetary boundaries: exploring the safe operating space for humanity. *Ecology and Society*, *14*(2). https://www.jstor.org/stable/26268316

Rogelj, J., Den Elzen, M., Höhne, N., Fransen, T., Fekete, H., Winkler, H., … & Meinshausen, M. (2016). Paris Agreement climate proposals need a boost to keep warming well below 2 C. *Nature*, *534*(7609), 631–639. https://doi.org/10.1038/nature18307

Schiermeier, Q. (2018). Droughts, heatwaves and floods: How to tell when climate change is to blame. *Nature*, *560*(7716), 20–22. https://doi.org/10.1038/d41586-018-05849-9

Schuelke-Leech, B. A. (2021). Disruptive technologies for a green new deal. *Current Opinion in Environmental Science & Health*, *21*, 100245. https://doi.org/10.1016/j.coesh.2021.100245

Seubelt, N., Michalke, A., & Gaugler, T. (2022). Influencing factors for sustainable dietary transformation—A case study of German food consumption. *Foods*, *11*(2), 227. https://doi.org/10.3390/foods11020227

Sithey, G., Thow, A.-M., & Li, M. (2015). Gross national happiness and health: Lessons from Bhutan. *Bulletin of the World Health Organization*, *93*(8), 514–514. https://www.scielosp.org/pdf/bwho/2015.v93n8/514-514/en

Spence, A., Poortinga, W., & Pidgeon, N. (2011). The psychological distance of climate change. *Risk Analysis*, *32*(6), 957–972. https://doi.org/10.1111/j.1539-6924.2011.01695.x

Steffen, W., Richardson, K., Rockström, J., Cornell, S. E., Fetzer, I., Bennett, E. M., Biggs, R., Carpenter, S. R., Vries, W., Wit, C. A., Folke, C., Gerten, D., Heinke, J., Mace, G. M., Persson, L. M., Ramanathan, V., Reyers, B., & Sörlin, S. (2015). Planetary boundaries: Guiding human development on a changing planet. *Science*, *347*(6223). https://doi.org/10.1126/science.1259855

Sustainable Food Trust. (2019). *The hidden cost of UK food.* Available at: https://sustainablefoodtrust.org/wp-content/uploads/2022/01/Website-Version-The-Hidden-Cost-of-UK-Food_compressed.pdf

TEEB. (2018). *Measuring what matters in agriculture and food systems. A synthesis of the results and recommendations of TEEB for agriculture and food's scientific and economic foundations report.* UN Environment. Available under: http://teebweb.org/agrifood/wp-content/uploads/2018/10/Layout_synthesis_sept.pdf

Thornton, P. K., Ericksen, P. J., Herrero, M., & Challinor, A. J. (2014). Climate variability and vulnerability to climate change: A review. *Global Change Biology, 20*(11), 3313–3328. https://doi.org/10.1111/gcb.12581

Tilman, D., Balzer, C., Hill, J., & Befort, B. L. (2011). Global food demand and the sustainable intensification of agriculture. *Proceedings of the National Academy of Sciences, 108*(50), 20260–20264. https://doi.org/10.1073/pnas.1116437108

True Cost Initiative. (2022). *TCA Handbook – Practical True Cost Accounting guidelines for the food and farming sector on impact measurement, valuation and reporting.* Available at: http://tca2f.org/wp-content/uploads/2022/03/TCA_Agrifood_Handbook.pdf

Tulder, R. V., Rodrigues, S. B., Mirza, H., & Sexsmith, K. (2021). The UN's sustainable development goals: Can multinational enterprises lead the Decade of Action? *Journal of International Business Policy, 4*(1), 1–21. https://doi.org/10.1057/s42214-020-00095-1

UN. (2015). *Transforming our world: The 2030 agenda for sustainable development.*

UN. (2016). *The Sustainable Development Goals Report 2016.* e-ISBN: 978-92-1-058259-9

UN. (2019). *Sustainable development goals—Decade of Action.* Available at: https://www.un.org/sustainabledevelopment/decade-of-action/

Unerman, J., Bebbington, J., & O'dwyer, B. (2018). Corporate reporting and accounting for externalities. *Accounting and Business Research, 48*(5), 497–522. https://doi.org/10.1080/00014788.2018.1470155

Weber, H., & Sciubba, J. D. (2018). The effect of population growth on the environment: Evidence from European regions. *European Journal of Population, 35*(2), 379–402. https://doi.org/10.1007/s10680-018-9486-0

Wiedmann, T., Lenzen, M., Keyßer, L. T., & Steinberger, J. K. (2020). Scientists' warning on affluence. *Nature Communications, 11*(1). https://doi.org/10.1038/s41467-020-16941-y

Wijkman, A., & Rockström, J. (2013). *Bankrupting Nature.* Routledge. https://doi.org/10.4324/9780203107980

World Economic Forum. (2022). *The global risks report 2022, 17th edition.* Available at: https://www3.weforum.org/docs/WEF_The_Global_Risks_Report_2022.pdf (Accessed November 13, 2022)

Part II

Theoretical Background and Development

2 Development of True Cost Accounting

Abstract
This section delves into the development of True Cost Accounting (TCA) and provides a comprehensive theoretical and economic foundation. A comparison between planned economy and market economy is made, highlighting the superiority of the latter despite its weaknesses. The section emphasizes the necessity of government interventions to address market failures, especially concerning environmental externalities. It explains how external costs can be integrated into product prices to reflect true prices and advocates for the introduction of Pigou taxes to internalize environmental costs. Additionally, the Impact-Weighted Accounts (IWA) Framework is introduced, aiming to quantify and monetize the broader impacts of business activities on stakeholders and the environment, thereby promoting a transition to a more sustainable and transparent economic model.

Key takeaways

After reading this chapter, you will:

- Understand the superiority of the market economy despite weaknesses
- Recognize the necessity of government interventions for addressing market failures, especially regarding environmental externalities
- Appreciate the importance of integrating external costs into product prices to reflect true values
- Advocate for Pigou taxes as a means to internalize environmental costs and promote sustainability
- Grasp the significance of the Impact-Weighted Accounts (IWA) Framework and other approaches for quantifying broader business impacts and fostering transparency

2.1 Theory-based classification and economic foundations

Key issue: What are the main principles of the market economy and why should further criteria (such as social responsibility and justice) be considered in addition to economic criteria to assure welfare-maximizing outcomes?

DOI: 10.4324/9781003345190-4

Generally speaking, companies are economic entities that combine different factors of production to create and sell goods and services (Wöhe et al. 2013, p. 36). Every company is confronted with largely system-independent challenges. These include (a) the need to procure factors of production, (b) the principle of profitability, and (c) the pursuit of financial equilibrium. In addition, system-dependent factors like taxes or other regulatory requirements resulting from a given economic system can also be identified.

There are clear and far-reaching implications with regard to the two – in many respects contrasting – economic systems (I) planned economy and (II) market economy. In the former, the institutions and decisions of the state play a central role: The state plans which goods are to be produced in which quantities and also determines the price of the goods (principle of plan fulfilment). Companies act as organs of the state and do not have the freedom to decide on production and sales conditions (principle of a central authority). They are mostly also state-owned (principle of common ownership). In contrast, companies in a market economy are geared to making a profit (principle of profit-making). In this context, companies define their own business plan independently. This implies that prices – for factors of production as well as for produced goods – are generated decentral on free markets (principle of autonomy). In addition, production resources are owned by the party providing the necessary capital. Companies are mostly privately owned (according to Wöhe et al. 2013, p. 36). Thus, in a planned economy the central decisions are made by the state, which maintains control of everything. In contrast, the market economy approach relies on independent and strong companies competing with each other via free markets (Schlecht 1990, Gutenberg 2013). Clearly, the ideal type of a free market economy is itself an ideologically charged oversimplifications. In reality, so-called free market economies are always dependent on state institutions (e.g., central banks, cartel offices, or the welfare state) whose activity can have an immense impact on the price formations of markets. In this sense, one could more adequately speak of different degrees of planned economies and attest from the historical collapse of the Soviet Union that an overly strong political control over market outcomes is economically inefficient.

Nevertheless, this did not result in the expected "end of history" (Fukuyama 1992). It is undisputed that market-economy and liberal-democratic forces gained global dominance in the following decades (Gowan 1999, Baker 2005) and paved the way for positive economic development. However, developments such as the bursting of the dotcom bubble (2000) (Goodnight and Green 2010) and the global financial and economic crisis (from 2008 onwards) (Helleiner 2011) clearly show that even an ideology strongly centered on market and economic forces is susceptible to crises or – to a greater or lesser extent – may even have caused them in the first place (Greenspan 2008). Massive government intervention was needed to successfully overcome both of these crises (Aikins 2009). Measures to control the COVID-19 pandemic (from 2019 onwards) (Ciotti et al. 2020) also led to (at times very strict) legislative policies, which were supported, and even welcomed, by the vast majority of the population (World Economic Forum 2020). The climate crisis has been looming for decades and is now becoming increasingly virulent in its consequences (IPCC 2023). It brings us to the issue whether a (neoliberal) market economy based on the premise of free markets can adequately address the causes of the global rise in temperatures and its ecological and social consequences (Fremstad and Paul 2022). We argue that global warming and its dramatic ecological and social consequences largely result from blinding out the basic economic theory of market failure. We shall go on to explain this theory and its economic efficiency in more detail and thus – on the basis of economic theory – derive a normative claim for state intervention. This will result in comprehensive requirements for those companies that are currently the largest distributors of greenhouse gases and other pollutants (IPCC 2001).

Before we look at the market failure of global scale and introduce ways of resolving it, we first need a fundamental understanding of the history and functioning of markets. The roots of today's economy are often located in Ancient Greece, where the term economy (from oikonomía, i.e., management of a household or state housekeeping) originates. The view of man as zõon politikón, i.e., as a person who lives in the community of a polis or city, dates back to the Greek philosopher and natural scientist Aristotle (384–322 BC) (Barker and Stalley 1995). At that time, there was not yet any differentiation between ethics, politics, and economics, but parts of practical philosophy centered around the question if the "good life" (Crespo 2014) was in the best interest of the community (Ricoeur 1990). Early modern economics reveals a more differentiated picture, with the Scottish moral philosopher and economist Adam Smith (1723–1790) co-founding the classic national economy. He proclaimed the existence of an invisible hand of the market (Rothschild 1994, Bishop 1995, Smith 1776, 2007). According to Smith, this organizing and regulating force of the market ensures that individual market participants pursue their own interests and thus contribute to the prosperity of society as a whole in the best possible way (Rothschild 1994). Against this backdrop, it has been variously argued that state intervention in principle is harmful and, with a few exceptions, should therefore be rejected. The role of the state should be limited to the provision of public goods, the limitation of monopolies, and the protection of property (Hayek 1944, Kurer 1989).

The supply and demand mechanism is fundamental for understanding why significant state intervention in the market mechanism can be necessary. Therefore, it makes sense that we look at the supply and demand mechanism in more detail.

Under the assumption that markets are perfect, with no company dominating the market (atomistic competition) and all market players being informed and able to act independently and immediately, without barriers to entry or exit from the market, and with fully defined property rights (Mankiw and

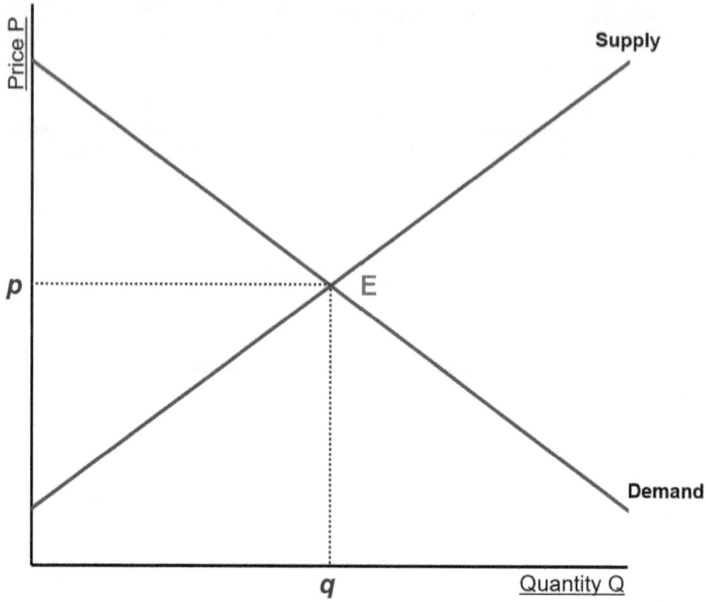

Figure 2.1 Basic model

Taylor 2021), partial markets deliver an efficient result, for homogeneous (i.e., identical) goods – and for the market as a whole. In an economic sense, this can be considered as a "correct" result.

This "correct" result can best be derived graphically using the basic model of supply and demand. In a price-quantity diagram referring to a partial market, i.e., the market for a particular commodity, the abscissa refers to the quantity (Q) of the commodity and the ordinate to its price (P) (see Figure 2.1).

The demand function results from the maximum willingness to pay of all consumers, shown in descending order. The resulting demand function therefore has a declining trend. To simplify matters, linearity is also assumed. In contrast, the supply curve is the aggregate of all available goods and services, arranged in ascending price order. Differences in supplier prices result from the fact that companies are able to offer the same product at different prices. This may be due to the use of different production technologies or the fact that economies of scale and scope come into play to varying degrees. Linearity is also assumed here to simplify the model. At the intersection of the two curves, the market is in equilibrium, which is indicated in the graph as point E (see Figure 2.1). This equilibrium relates both to the quantity disposed of at the point q at the intersection with the abscissa, and to the market price, which can be determined at point p at the intersection with the ordinate.

Consumers with a lower maximum willingness to pay than the price p established on the market are excluded, as are suppliers who can only offer the product at a price above p.

On the other hand, all consumers for whom the product or service is worth more than p and all producers who can offer it for less than p can participate in the market. The resulting market exchange is advantageous for both sides and, in abstract terms, is referred to as a welfare gain. This is made up of an overall benefit to the demand side, which is called the consumer surplus (CS) and can be understood as the area between the ordinate axis, the demand curve, and the straight line through p and E. The respective consumers thus benefit from the market price being below their willingness to pay, represented by the demand curve. A welfare gain also consists of the overall benefit to the supplier, which is referred to as the producer surplus (PS). This can be visualized as the area between the x-axis, the supply curve, and the straight line through x and E (see Figure 2.2).

The market equilibrium presented here is Pareto-optimal, i.e., no market participant can be made better off without making another market participant worse off. The sum of CS and PS is thus at a maximum, resulting in the highest possible welfare benefit to society as a whole.

If one or more of the above assumptions are compromised, the market is no longer perfect and the maximum welfare gain outlined above decreases. This is particularly the case when natural goods such as a clean atmosphere are available as a public commodity. In this case, the two typical criteria of

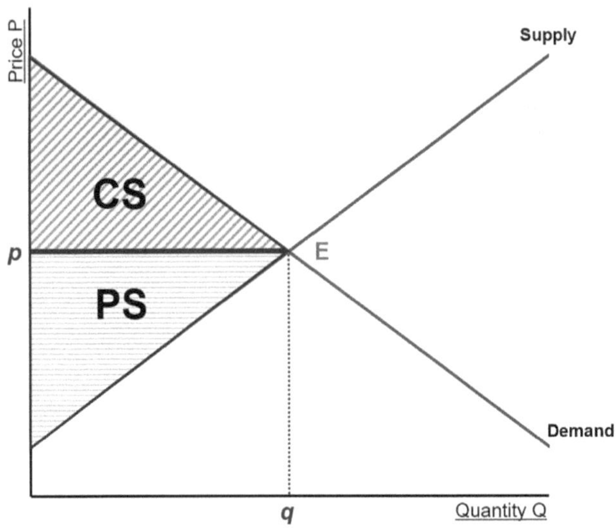

Figure 2.2 Welfare equilibrium

"non-rivalry in consumption" and "non-excludability" apply as there are no fully defined ownership rights. Every market participant benefits from a clean atmosphere but, at the same time, can pollute it at little or no cost. Greenhouse gas emissions in the atmosphere illustrate this dilemma. Any market participant can emit greenhouse gases without limitation. However, the initially clean atmosphere has to absorb the unbridled emissions of all market participants and ultimately becomes an overused carbon sink. This, in turn, has negative effects on all parties involved, which we perceive – in this example – as climate change and its consequences (IPCC 2013). As a result, public goods are overused, which has a negative impact on overall welfare.

Let's now look at this relationship in more detail. If the production of a commodity results in a polluted atmosphere, the resulting costs to society would have to be added to the price. Its price would have to rise from the price p derived at the beginning to the price p' (see Figure 2.3). At this higher price, however, there would not be enough demand. Initially there would be an excess supply until the new equilibrium price p* emerges at the intersection between the new, more expensive (i.e., shifted upwards to the left) supply function and the unchanged demand function. This is accompanied by a reduced demand for the commodity, which drops from q to q*. In short, including the actual costs of air pollution increases the price of the commodity by Δp. Thus, the quantity demanded Δq decreases in the new equilibrium.

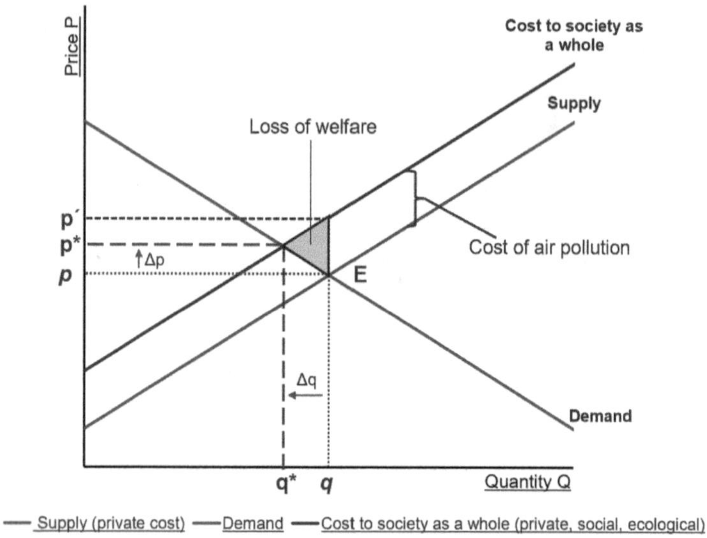

Figure 2.3 Welfare gained

To ensure that society as a whole is supplied with a public good (such as a clean atmosphere, as shown in our example), the responsible parties (in our example the emitters of greenhouse gases) must include the costs of air pollution in their supply function. Otherwise, there would be a loss of welfare equivalent to the triangle highlighted in Figure 2.3. To be economically efficient and maximize welfare, the emitter (i.e., the company placing the product on the market) should be held directly responsible for their environmentally damaging behavior. This approach is also in line with the polluter pays principle, which was agreed by the UN back in 1992 (Schwartz 2010).

Two approaches appear to be conceivable to move from the status quo to the socially optimal solution. In the first approach, the parties directly involved negotiate a solution. This procedure can be illustrated by a simple example. A cattle farmer worries about the health of his herd because the neighboring farmer treats his wheat fields with pesticides. Both farmers depend on each other. The cattle farmer needs the wheat as feed for his animals. The wheat farmer needs the dung of the cattle as fertilizer. The two farmers could then seek a negotiated solution with the aim of maximizing joint profits. As a result, the wheat farmer would use less pesticides and thus ensure the well-being of the cattle farmer (Vogel 1987). As Coase (1960) was able to demonstrate, it is possible – directly at the level of the market participants concerned and without state intervention – to eliminate negative external effects efficiently (Wenzel 2022), meaning the negotiated solution would not result in any loss of welfare. However, the **Coase theorem** only applies under the following very restrictive assumptions: There are no transaction costs, i.e., there are no costs associated with negotiations between the two parties; the two parties also have access to complete information, including on the costs to both parties for prevention and damage; all ownership rights are clearly distributed, regardless of the original arrangements for the rights (Harrison and McKee 1985). However, these assumptions are not in line with reality. Transaction costs are always present, at the very least parties have to invest time and personnel in the negotiation. Also, complete information on prevention and damage costs is almost never available. Ownership rights can also tend to be controversial in real-life scenarios. The simplistic example provided also seems unrealistic, as the circle of parties involved and affected will in reality go far beyond just two contracting parties (Aivazian and Callen 1981, Hoffman and Spitzer 1986). This is particularly evident in the case of atmospheric pollution. Here, the effects can be seen on a global scale. In this context (and taking historical observations into account), it is clear that this approach cannot be applied to any significant extent in reality (Aivazian and Callen 2003).

The second approach involves the state as an active regulator. This requires the emitter to include in its calculation the actual environmental and social costs associated with its production, i.e., to internalize them. Adding such costs to the prize of a commodity is known as a **Pigovian tax**. As a result, the goods produced would have to be offered at a higher price, which would lead to a fall in demand. This would have negative consequences for the producer and probably also be perceived as negative by customers. Nevertheless, if the responsible party would integrate the externalities, welfare for society as a whole would increase in the magnitude of the triangle highlighted in Figure 2.3. However, the main focus of the Pigovian tax is not on the fiscal effect of a tax (i.e., the increase in government revenues), but on the avoidance of environmental pollution (Kallbekken 2013).

It could be argued that it is difficult to accurately determine the costs resulting, for example, from the emission of greenhouse gases, and that these will also change over time. Although this argument is fundamentally correct, the criteria of social responsibility and justice should be considered in addition to economic criteria regardless of their complexity. The status quo places the burden of the resulting damage in particular on both the financially weak Global South and on future generations by not adhering to all relevant criteria. According to consumer opinions, politicians should be held responsible for taking action and eliminating market failures by involving the scientific community and the respective parties (Michalke et al. 2022, Semken et al. 2024).

For the sake of completeness, it should be mentioned that the state, as a regulatory actor, has other measures at its disposal, such as bans or the introduction of maximum emission levels for individual companies. However, these options, which are part of standard regulatory law, are not discussed in more detail below, because they cannot, taken on their own, claim to lead to efficient, i.e., welfare-maximizing solutions (Rave et al. 2013).

2.2 External effects and their implications at company level

Key issue: What are the implications of an internalization of externalities at the company and product level?

We shall now systematically address the implications of internalizing externalities (the polluter pays principle). Let us start by considering a socially and ecologically unsustainable (further mentions of "un/sustainability" in this chapter, if not otherwise stated always refers to only the social and ecological dimension of sustainability) company that offers an exemplary product at the current (market) price of EUR 100. If the negative external effects, in the form of the high environmental and social damage caused by this company, were added to the current price, this would result in a higher price. Let us assume, this higher price already includes any positive externalities on society or nature, which are positive effects that benefit society rather than the company, such as the planting of trees (ecosystem services) (see Figure 2.4a). In contrast, a sustainably operating company produces the same product as the unsustainable company described above but in compliance with high environmental and social standards. This company is currently able to offer the product at a current (market) price of EUR 125. Therefore, they have difficulties

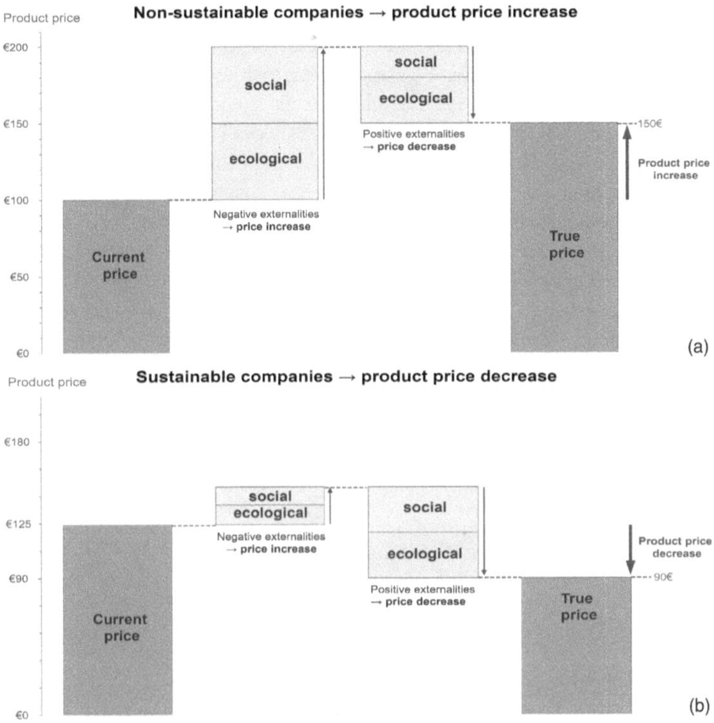

Figure 2.4 Comparison between actual prices and true prices in unsustainable (a) and sustainable companies (b), taking negative and positive ecological and social externalities into account. The product price of unsustainable companies increases, while the product price of sustainable companies decreases

finding buyers for their (more expensive) product. They may only be able to serve a rather small niche market. Their negative environmental and social externalities turn out to be relatively low; their positive externalities (ecosystem services) are comparatively high. If all externalities for which the sustainable manufacturing company is responsible were to be internalized according to the originator, its product price would fall and drop to an exemplary value of EUR 90 (see Figure 2.4b).

Comparing the two companies shows that the unsustainable company, which was previously able to offer its product at a lower price than the sustainable company (albeit at the expense of the environment and with negative social implications) is no longer able to do so when the externalities from production are taken into account. On the contrary, its unsustainable product – measured with true prices – is now even more expensive than that of its sustainably operating competitor.

If all externalities were internalized, an absolute price variation (i.e., a price reversal) would occur for a large number of products, as presented in this example. However, it can be assumed that there will often be no absolute price variation, but only a relative one, especially due to economies of scale and the often poor economic performance of socially and ecologically sustainable products. This means that the products of unsustainable companies are cheaper than their counterparts produced by sustainable companies, even at true prices. Nevertheless, the price difference between the two products would narrow in favor of the sustainable company.

If external effects were integrated into pricing, this would lead to a significant strengthening of sustainable companies. On the other hand, integrating external effects would create strong incentives for all companies, even those that are already sustainable today, to reduce negative externalities to a much greater extent and to accelerate the provision of positive externalities.

2.3 Scientific background of True Cost Accounting

Key issue: How can True Cost Accounting be implemented and what are initiatives adapting True Cost Accounting?

2.3.1 The Impact-Weighted Accounts Framework

Key issue: What is the Impact-Weighted Accounts Framework?

The Impact-Weighted Accounts (IWA) Framework was developed under the leadership of the nonprofit Impact Economy Foundation, Amsterdam, and was published ongoingly between June 2022 and February 2023 (IEF 2022). Development of the framework was supported in particular by the Harvard Business School, Singapore Management University, Rotterdam School of Management, Erasmus University and the Impact Institute, which was established in 2018 as a spin-off of the social enterprise True Price (IEF 2022, True Price 2023). In deliberate contrast to Milton Friedman's doctrine that "the social responsibility of business is to increase its profits" (Friedman 1970), the Impact Economy Foundation stipulates that "[t]he purpose of business is to create true value" (IEF 2022). While the current economy (unregulated capitalism) with extractive business models aims to increase financial value alone, the authors want to contribute to an "impact economy" that is characterized by generative business models and want to make a move away from traditional drivers/target variables such as (short-term oriented) profits, prices and returns to true profits, true prices, and true returns. To achieve this, the proposal is to introduce an Integrated Profit and Loss (IP&L) statement, which initially supplements a traditional Profit and Loss statement with the financial impact on stakeholder groups, defined here as customers, suppliers, employees, and the broader society. In a further step, it will then be determined what broader impact the company under investigation has on its stakeholders (over one year). The framework aims to quantify and monetize all impacts. To this end, the IP&L statement is divided into six different types of capital: First, three types of economy-related capital, namely (1) financial (e.g., payments to suppliers, payments from clients, wages, profits and taxes), (2) manufactured (e.g., produced products and change in fixed assets), and (3) intellectual capital (e.g., intellectual property and technological development). Second, two types of company-related capital, namely (4) social (e.g., social cohesion, contribution to human rights, child labor) and (5) human capital, which – more narrowly defined than human capital – refers to the value of time invested by employees to work, the well-being of employees and customers and their safety, as well as ecology-related (6) natural capital (e.g., contribution to climate change, biodiversity as well as soil, water and air pollution) (IEF 2022). To determine the IP&L statement, the Impact Economy Foundation proposes a five-stage process: After the data to be collected are identified, measured, and monetized, they are aggregated within the six different types of capital and finally presented (IEF 2022). The framework is intended to enable an outward-looking, comprehensive accountability that allows potential investors and the capital market to correctly assess a company's actual value creation and thus obtain a valid picture of its long-term risks and its long-term financial viability. Another objective of the framework is to allow an internal management system that aims to improve the impact of the company.

In addition to the introduction of an IP&L statement, the Impact Economy Foundation is proposing a second approach (IEF 2022) still under development, namely the creation of an Integrated Balance Sheet (IBaS). This will expand the existing financial balance sheet by including the six types of capital allocated to assets and liabilities. Thereby, the IBaS provides comprehensive information on a company's situation at the end of the financial year. While it is unlikely that the IBaS will become common practice soon, at the very least the traditional balance sheet should be expanded to include natural capital, which is to be found on the liability side, as well as the integration of human capital on the assets side and social capital, which is to be included on both sides of the balance sheet (IEF 2022).

2.3.2 Economics of Ecosystems & Biodiversity (TEEB) and the Capitals Coalition

Key issue: What are the Economics of Ecosystems and Biodiversity (TEEB) and the Capitals Coalition and how are they related?

The work of the Capitals Coalition can be traced back to the Economics of Ecosystems & Biodiversity (TEEB) research initiative. TEEB was launched in 2007 by Germany and the European Commission with the aim of quantifying the value of the ecosystem services provided by nature, as well as the value of biodiversity, and expressing them in monetary terms. This has resulted in several publications, the latest one a synthesis report published in October 2010 (TEEB 2010). In the following years, researchers continued working on natural capital accounting and further differentiated the methodology for determining the value of the ecosystem and its services. They also implemented this methodology in pilot projects in various countries outside the EU. The follow-up program TEEBAgriFood was supported by the United Nations Environment Programme. It was launched specifically for the agricultural sector, the results of which were published in 2018 (TEEB 2018a, 2018b).

The TEEB for Business Coalition was founded in 2012. It is aimed at further developing the insights of the TEEB initiative for companies. TEEB for Business Coalition formed the Natural Capital Coalition in 2014, which is now known as the Capitals Coalition (2023a). The Capitals Coalition operates as a global not-for-profit platform (Capitals Coalition 2023a) and is supported by a large number of companies such as ABN AMRO, L'Oreal, and Novartis, as well as NGOs such as WWF International (Capitals Coalition 2021a). The Capitals Coalition distinguishes between four types of capital: (1) Natural capital covers the ecology dimension, i.e., the available natural resources. At the level of society, they make a distinction between (2) social capital and (3) human capital. They define social capital along the OECD (2001, p. 41) as "networks together with shared norms, values and understanding that facilitate cooperation within and amongst groups". They define human capital according to Keeley (2007) as representing the knowledge and ability of individuals to create well-being on a personal, social, and economic level. The (4) produced capital is the only dimension that can already be extensively monetized today (Capitals Coalition 2021a, 2023b).

In January 2021, the Capitals Coalition published two protocols to help companies close the gap in the evaluation of the first three types of capital: First, the Natural Capital Protocol (Capitals Coalition 2021b), which focuses on the environmental dimension; second, the Social & Human Capital Protocol (Capitals Coalition 2021a), which addresses the social dimension. In the words of the Capitals Coalition, the two protocols are "decision-making frameworks that enable organisations to identify, measure and value their impacts and dependencies on natural capital, social capital and human capital" (Capitals Coalition 2023b). To achieve this, they propose a four-stage approach. Initially, the company should clarify why a natural capital assessment should be carried ("frame" stage). Next, the company must establish what is to be achieved with the help of the assessment ("scope" stage). In the third ("measure and value") stage, the company focuses primarily on methodological questions. This includes measuring and evaluating the drivers of ecological, social, and human capital. In the fourth ("apply") stage, the company identifies how the results can be used meaningfully to make targeted changes to existing processes.

The proposals presented by the Capitals Coalition in both protocols aim to identify the negative impacts of businesses on the environment and society so that these can subsequently be reduced and positive impacts can be increased. Overall, the two protocols have a generic character in that they offer users a comprehensive overview of methods with which companies can quantify and monetize their environmental and social impacts. Occasionally, exemplary breakthroughs made by individual companies are listed in these protocols (Capitals Coalition 2021a, 2021b). In addition, sector-specific guides are provided by the Capitals Coalition (2016), which were published some time before the two protocols. They focus on the three product-related sectors of apparel (Capitals Coalition 2016), forest products, and food and beverage. They also cover biodiversity and the finance sector (Capitals Coalition 2018, 2024).

The Capitals Coalition integrates the company's fields of action into an "infinity loop of interactive relationships" (Goodrich et al. 2022, p. 11), which – in addition to the manufacturing economy ("business") – also includes the "finance" (service) sector and politics ("governance"). This loop aims to show the roles and relationships between all three stakeholders in order to create a "more inclusive form of capitalism" (Goodrich et al. 2022, p. 11). In close cooperation between business, finance, and governance, the aim is to work toward a capitalism "that generates value for nature, people and society alongside business and traditional economies" (Goodrich et al. 2022, p. 11); a goal that is strikingly similar to that of the Impact Economy Foundation and the IWA Framework described above, despite different organizational backgrounds and different procedural approaches.

2.3.3 Sustainable Performance Accounting

Key issue: What is Sustainable Performance Accounting?

Founded in 2006, Regionalwert AG Freiburg allows citizens to actively take responsibility for the region and sustainable agriculture through the purchase of shares (Regionalwert AG 2024a). In the subsequent years, founder Christian Hiß developed the proposal to consider operational sustainability performance in corporate financial reporting. From this idea emerged the methodology of Sustainable Performance Accounting (SPA) for corporate accounting, which aims to integrate the sustainability potentials and risks of a company into conventional profit and loss statements and balance sheets (Regionalwert AG 2024b). The think tank Regionalwert Research gGmbH that was established in 2022 continues this work. In collaboration with Regionalwert Leistungen GmbH and the University of Emden-Leer they published a detailed description of the SPA methodology (Henkel et al. 2023).

SPA aims to incorporate all operational risks, particularly environmental, social, and governance (ESG) risks and performances, into the balance sheet and the profit and loss statement to achieve integrated sustainability accounting (Lay-Kumar et.al. 2023). This approach elevates sustainability performance and costs to the same level as financial metrics. This stems from the understanding that all economic activities are embedded in socio-ecological contexts and that ESG performances, which have not been traditionally included in financial success metrics, are central components of corporate success (Henkel et al. 2023).

In practice, SPA integrates sustainability risks or potentials into accounting by recording ESG expenditures or potentials in a secondary ledger, known as the ESG ledger (Liu et al. 2024). The sum of conventional accounting and ESG accounting then forms the Sustainability Accounting. This provides a picture that accurately reflects the true financial, asset, and earnings situation of the company. For example, CO_2 emissions are represented by the creation of a CO_2 provision in the ESG ledger, while CO_2 sequestration is internalized by the activation of ESG assets (Henkel et al. 2023). Therefore, previously nonfinancial assets are replaced by pre-financials, which are characterized by both financial materiality and impact materiality (IDW 2020, Henkel et al. 2023). This indicates that, in the medium term, they will exert an influence on the creditworthiness and the value of the company. With a stringent disclosure requirement for all operational risks, especially social and ecological risks, which currently only appear in the ESG balance sheet, these risks would also be reflected in the conventional balance sheet. Even though the ESG reporting developed in SPA currently goes significantly beyond the Corporate Sustainability Reporting Directive (CSRD) requirements (Henkel et al. 2023), it is anticipated that the ESG accounting developed within SPA will gain significant importance in the future. Already in the fiscal year 2024, the first German companies will have to prepare a sustainability report in accordance with the CSRD. Similar measures are likely to be added as the EU pursues its Green Deal goal of becoming climate neutral by 2050 (EUCOM 2019), SPA demonstrates that ESG issues can be integrated into real performance metrics through ESG accounting. As this is currently only possible through the additional effort of maintaining a separate ESG ledger, it is necessary for more ESG matters to be integrated into regular accounting, rendering the separate ESG accounting redundant. However, since there are still no valuation methods for many sustainability issues, the discourse must initially focus on developing robust indicators for measuring sustainability.

2.3.4 *Auditing and accounting firms view on True Cost Accounting*

Key issue: How do auditing and accounting firms view and apply True Cost Accounting?

Several auditing and consulting firms are deeply engaged with True Cost Accounting (TCA). The Boston Consulting Group (BCG) notes that external costs are often not borne by the polluter but by society, leading to market distortions (Boston Consulting Group 2020). In their report "Securing the Future of German Agriculture" (Boston Consulting Group 2020), BCG examines the critical challenges and necessary reforms to ensure the sustainability and resilience of German agriculture. The report underscores the need for a balance between economic viability and environmental sustainability, urging the adoption of innovative agricultural technologies and practices to enhance resource efficiency. It highlights the importance of aligning with the EU Green Deal's objectives, which aim to make the European economy climate-neutral by 2050.

Key recommendations from BCG include the implementation of digital technologies, precision farming, and sustainable land management practices to improve productivity while reducing environmental impact. The report stresses the need for robust policy frameworks that provide financial incentives for sustainable farming practices and support research and development in agricultural innovation. Collaboration across the supply chain is identified as essential for driving the transition toward sustainable agriculture. This involves partnerships between farmers, policymakers, businesses, and consumers to promote and implement eco-friendly practices. Additionally, the report calls for increased public awareness and education on the importance of sustainable agriculture and its benefits to society and the environment. By addressing these recommendations, the report suggests that German agriculture can achieve a more sustainable and resilient future, ensuring food security, protecting natural resources, and contributing to the broader goals of climate neutrality and environmental stewardship.

Other accounting firms like Ernst & Young (EY) also support TCA as a method to incorporate the hidden environmental and social costs of food. EY, for example, emphasizes that TCA can help transform the food system, ensuring that prices reflect the real impact on nature and society, thus driving long-term value creation and reducing the burden on future generations (Lugtigheid 2023). KPMG Group sustains a "True Value" methodology, which helps companies to measure and monetize the positive and negative economic, social, and environmental externalities of their business activities. This approach enhances strategic decision-making, transparency, and stakeholder relations (Krause 2024).

To conclude, auditing and consulting firms such as BCG, EY, and KPMG advocate for TCA, emphasizing the need to incorporate hidden environmental and social costs into business practices to promote sustainability, align with global climate goals, and drive long-term value creation.

References

Aikins, S. (2009). Global financial crisis and government intervention: A case for effective regulatory governance. *International Public Management Review*, *10*(2), 23–43.

Aivazian, V. A., & Callen, J. L. (1981). The Coase theorem and the empty core. *The Journal of Law and Economics*, *24*(1), 175–181.

Aivazian, V. A., & Callen, J. L. (2003). The core, transaction costs, and the Coase theorem. *Constitutional Political Economy*, *14*(4), 287–299. https://doi.org/10.1023/B:COPE.0000003859.10184.f3

Baker, R. W. (2005). *Capitalism's Achilles Heel: Dirty Money and How to Renew the Free-Market System*. John Wiley & Sons.

Barker, E., & Stalley, R. F. (1995). *Aristotle, Politics*. Oxford University Press.

Bishop, J. D. (1995). Adam Smith's invisible hand argument. *Journal of Business Ethics, 14*, 165–180.

Boston Consulting Group. (2020). *Die Zukunft der deutschen Landwirtschaft sichern: Innovationen und Investitionen für eine nachhaltige Zukunft*. Available at: https://web-assets.bcg.com/7a/17/971c6d0e4fcb8067d406b8a9bb4a/die-zukunft-der-deutschen-landwirtschaft-sichern.pdf

Capitals Coalition. (2016). *Apparel sector guide*. Available at: https://capitalscoalition.org/guide_supplement/apparel-sector-guide/

Capitals Coalition. (2018). *Forest products sector guides*. Available at: https://capitalscoalition.org/guide_supplement/forest-products-sector-guides/

Capitals Coalition. (2021a). *Social and human capital protocol*. Available at: https://capitalscoalition.org/wp-content/uploads/2021/01/Social_and_Human_Capital_Protocol.pdf

Capitals Coalition. (2021b). *Natural capital protocol*. Available at: https://capitalscoalition.org/wp-content/uploads/2021/01/NCC_Protocol.pdf

Capitals Coalition. (2023a). *Green economy coalition*. Available at: https://www.greeneconomycoalition.org/members/capitals-coalition

Capitals Coalition. (2023b). *The capitals approach*. Available at: https://capitalscoalition.org/capitals-approach/

Capitals Coalition. (2024). *Guides & supplements*. Available at: https://capitalscoalition.org/capitals-approach/guides-and-supplements/

Ciotti, M., Ciccozzi, M., Terrinoni, A., Jiang, W.-C., Wang, C.-B., & Bernardini, S. (2020). The COVID-19 pandemic. *Critical Reviews in Clinical Laboratory Sciences, 57*(6), 365–388.

Coase, R. H. (1960). The problem of social cost. *Journal of Law and Economics, 3*, 1–44.

Crespo, H. (2014). *Tending the Epicurean Garden*. Humanist Press.

Economics of Ecosystems and Biodiversity (TEEB). (2018a). *Measuring what matters in agriculture and food systems: A synthesis of the results and recommendations of TEEB for Agriculture and Food's Scientific and Economic Foundations report*. UN Environment., June 2018. ISBN: 978-92-807-3702-8. Available at: http://teebweb.org/agrifood/wp-content/uploads/2018/10/Layout_synthesis_sept.pdf

Economics of Ecosystems and Biodiversity (TEEB). (2018b). *TEEB for Agriculture & Food: Scientific and Economic Foundations*. UN Environment. Available at: https://teebweb.org/wp-content/uploads/2018/11/Foundations_Report_Final_October.pdf

European Commission (EUCOM). (2019). *Mitteilung der Kommission: Der europäische Grüne Deal (COM(2019) 640 final)*. Available at: https://eur-lex.europa.eu/legal-content/DE/TXT/?uri=COM:2019:640:FIN

Fremstad, A., & Paul, M. (2022). Neoliberalism and climate change: How the free-market myth has prevented climate action. *Ecological Economics, 197*, 107353.

Friedman, M. (1970). of Business is to increase its profits. *New York Times Magazine*, September, *13*, 122–126.

Fukuyama, F. (1992). *The End of History and the Last Man*. New York: The Free Press.

Goodnight, G. T., & Green, S. (2010). Rhetoric, risk, and markets: The dot-com bubble. *Quarterly Journal of Speech, 96*(2), 115–140.

Goodrich, R., **C. H.**, **Mohamed**, **N.**, **Lok**, **M.**, **Maughan**, **J.**, & **Masson**, **A. (2022).** Nature *positive economies: Progress, lessons and next frontiers*. Available at: https://www.greeneconomycoalition.org/news-and-resources/nature-positive-economies-progress-lessons-and-next-frontiers

Gowan, **P.** (**1999**). *The Global Gamble: Washington's Faustian Bid for World Dominance*. Verso.

Greenspan, **A. (2008).** We will never have a perfect model of risk. *Financial Times*, 16(3). Available at: https://www.ft.com/content/edbdbcf6-f360-11dc-b6bc-0000779fd2ac

Gutenberg, E. (**2013**). *Grundlagen der betriebswirtschaftslehre* (Vol. 2). Springer-Verlag.

Harrison, **G. W.**, & **McKee**, **M.** (**1985**). Experimental evaluation of the Coase theorem. *The Journal of Law and Economics*, *28*(3), 653–670.

Hayek, **F. A. (1944).** *The Road to Serfdom*. University of Chicago Press.

Helleiner, **E. (2011).** Understanding the 2007–2008 global financial crisis: Lessons for scholars of international political economy. *Annual Review of Political Science*, *14*, 67–87.

Henkel, **K.**, **Lay-Kumar**, **J.**, & **Hiß**, **C. (2023).** *Sustainable Performance Accounting (SPA) am Beispiel der Bilanzierung von CO2-Emissionen*. Hochschule Emden/Leer.

Hoffman, **E.**, & **Spitzer**, **M. L.** (**1986**). Experimental tests of the Coase theorem with large bargaining groups. *The Journal of Legal Studies*, *15*(1), 149–171.

IEF. (2022). *Guidance on the steps for compiling Impact-Weighted Accounts*. Impact Economy Foundation. Available at: https://impacteconomyfoundation.org/download/

Institut der Wirtschaftsprüfer (IDW). (2020). *IDW- Positionspapier "Sustainable Finance als Teil der nachhaltigen Transformation" - Auswirkungen auf Kreditinstitute* (13). IDW. Available at: https://www.idw.de/IDW/Medien/Positionspapier/Downloads-IDW/Down-Positionspapier-SustainableFinance.pdf

IPCC. (2001). *Climate Change 2001: The Scientific Basis. Contribution of Working Group I to the Third Assessment Report of the Intergovernmental Panel on Climate Change* (Houghton, J. T., Ding, Y., Griggs, D. J., Noguer, M., van der Linden, P. J., Dai, X., Maskell, K., & Johnson, C. A. (eds), 881 pp.). Cambridge University Press.

IPCC. (2013). *Climate change 2013: The physical science basis. Contribution of working group I to the fifth assessment report of the Intergovernmental Panel on Climate Change*. Cambridge University Press. Available at: https://www.ipcc.ch/site/assets/uploads/2017/09/WG1AR5_Frontmatter_FINAL.pdf

IPCC. (2023). Sections. In: Core Writing Team, Lee, H., & Romero J. (eds) *Climate Change 2023: Synthesis Report. Contribution of Working Groups I, II and III to the Sixth Assessment Report of the Intergovernmental Panel on Climate Change* (35–115). IPCC. https://doi.org/10.59327/IPCC/AR6-9789291691647. Available at: https://www.ipcc.ch/report/ar6/syr/downloads/report/IPCC_AR6_SYR_LongerReport.pdf

Kallbekken, S. (2013). Pigouvian Taxes and the Efficiency of Environmental policies. In: Shogren, J. F. (eds) *Encyclopedia of Energy, Natural Resource, and Environmental Economics*. Available at: https://www.sciencedirect.com/referencework/9780080964522/encyclopedia-of-energy-natural-resource-and-environmental-economics#book-info

Keeley, **B. (2007).** Human capital: A revolution?. *Organisation for Economic Cooperation and Development. The OECD Observer*, (261), 25.

Krause, **G. (2024).** *True value: More insight for right decisions*. KPMG. Available at: https://kpmg.com/de/en/home/insights/overview/more-insight-for-right-decisions/true-value.html

Kurer, O. (1989). John Stuart Mill on government intervention. *History of Political Thought*, *10*(3), 457–480. http://www.jstor.org/stable/44797145

Lay-Kumar, J., Walkiewicz, J., Hiß, C., & Heck, A. (2023). Sustainable Performance Accounting als Schlüssel für die Nachhaltigkeitstransformation – Einsichten aus dem Projekt QuartaVista. In: Zwick, Y. & Jeromin, K. (eds) *Mit Sustainable Finance Die Transformation Dynamisieren*. Springer Gabler. https://doi.org/10.1007/978-3-658-38044-1_14

Liu, X., Liang, W., Fu, Y., & Huang, G. Q. (2024). Dual environmental, social, and governance (ESG) index for corporate sustainability assessment using blockchain technology. *Sustainability*, *165*(10), 4272. https://doi.org/10.3390/su16104272

Lugtigheid, P. (2023). *True cost accounting: Wat is de werkelijke prijs van jouw appel?* EY. Available at: https://www.ey.com/nl_nl/sustainability/true-cost-accounting-wat-is-de-werkelijke-prijs-van-jouw-appel (Accessed July 9, 2024)

Mankiw, N. G., & Taylor, M. P. (2021). Grundzüge Der Volkswirtschaftslehre (8th ed.). Schäffer-Poeschel.

Michalke, A., Stein, L., Fichtner, R., Gaugler, T., & Stoll-Kleemann, S. (2022). True cost accounting in agri-food networks: A German case study on informational campaigning and responsible implementation. *Nachhaltigkeitswissenschaft*, *17*(6), 2269–2285.

OECD. (2001). *The well-being of nations, the role of human and social capital*. Available at: https://one.oecd.org/document/DEELSA/ELSA/ED/CERI/CD(2000)3/REV2/en/pdf

Rave, T., Triebswetter, U., & Wackerbauer, J. (2013). *Koordination von Innovations-, Energie-und Umweltpolitik* (No. 61). ifo Forschungsberichte.

Regionalwert AG. (2024a). *Über uns*. Available at: https://www.regionalwert-ag.de/informieren/ueber-uns/ (Accessed July 9, 2024)

Regionalwert AG. (2024b). *Sustainable Performance Accounting*. Available at: https://regionalwert-research.de/sustainable-performance-accounting/ (Accessed July 9, 2024)

Ricoeur, P. (1990). *Soi-même comme un autre*. Seuil.

Rothschild, E. (1994). Adam Smith and the invisible hand. *The American Economic Review*, *84*(2), 319–322.

Schlecht, O. (1990). *Grundlagen und Perspektiven der sozialen Marktwirtschaft* (Vol. 27). Mohr Siebeck.

Schwartz, P. (2010). The Polluter-Pays Principle. In: Fitzmaurice, M., Ong, D. M., & Merkouris, P. (eds) *Research Handbook on International Environmental Law*. Edward Elgar Publishing.

Semken, C., Allcott, H., Gaugler, T., Michalke, A., Stein, L., van Sambeek, F., Varela Seone, S., & Oebel, B. (2024). *Forthcoming - under review. Optimal green retailing: Theory and Evidence*. National Bureau of Economic Research.

Smith, A. (1776). An inquiry into the nature and causes of the wealth of nations. Available at: https://era.ed.ac.uk/handle/1842/1455

Smith, A. (2007). The Wealth of Nations: An inquiry into the nature and causes of the Wealth of Nations. With an introduction by Jonathan B. Wight, University of Richmond. Harriman House Limited. Publisher: Harriman House LTD. ISBN: 978-1-906659-87-5

TEEB. (2010). *The Economics of Ecosystems and Biodiversity: Mainstreaming the Economics of Nature: A synthesis of the approach, conclusions and recommendations of TEEB*. ISBN 978-3-9813410-3-4. Available at: https://www.teebweb.

org/wp-content/uploads/Study%20and%20Reports/Reports/Synthesis%20report/TEEB%20Synthesis%20Report%202010.pdf

True Price. (2023). *About True Price*. Available at: https://trueprice.org/about-us/

Vogel, K. R. (1987). The Coase theorem and California animal trespass law. *The Journal of Legal Studies*, *16*(1), 149–187.

Wenzel, M. (2022). Concluding Remarks: Meet the Future—with International Business Administration and Entrepreneurship. In Wenzel, M. (eds), *Foundations of Management & Entrepreneurship: Concepts and Cases* (1st ed., 387–390).

Wöhe, G., Bilstein, J., Ernst, D., & Häcker, J. (2013). *Grundzüge der Unternehmensfinanzierung*. Vahlen.

World Economic Forum. (2020). *The net-zero challenge: Fast-forward to decisive climate action*. Available at: https://www3.weforum.org/docs/WEF_The_Net_Zero_Challenge.pdf

3 Status Quo of True Cost Accounting

Abstract

This chapter explains the process of calculating true costs according to established methods. True Cost Accounting (TCA) comprises four steps: (1) Qualification of relevant ecological and social indicators, (2) quantification of indicators' values along the system's boundaries, (3) monetization and internalization of these values to derive the true costs, and (4) communication of results toward relevant stakeholders. We explain prevailing calculation methods and describe their uncertainties to transparently depict the potentials and burdens of TCA. Depending on the system's boundaries (e.g., one unit of product, business process, whole enterprise) the best calculation approach varies. Moreover, depending on the targeted audience, different ways to communicate the results of TCA are reasonable. Consequently, TCA has to be tailored to the users' needs. For a better understanding, in the following sections, each step of the TCA method is explored along one exemplary product for better understanding.

Key takeaways

After reading this chapter, you will:

- Be able to follow a state-of-the-art true cost calculation
- Gain insights into affiliate methods of life cycle assessment (LCA) and environmental costing
- Understand how environmental and social impacts are translated to economic values
- Acknowledge the difficulties and uncertainties arising with calculating environmental and social externalities
- Learn different ways of using True Cost Accounting (TCA) results for corporate activities or external communication

3.1 Accounting true costs: A method to monetize environmental and social impact

Key issue: How is TCA structured?

DOI: 10.4324/9781003345190-5

In this section, we explore the calculations that build the background of True Cost Accounting (TCA). With these calculations, environmental and social impacts from corporate activities are translated to economic terms by monetization. To provide the reader with a deeper understanding of the needed calculation steps, we perform them using an exemplary product.

Literature already provides a wide range of TCA approaches. For example, Ernst & Young (EY 2016) have published their Total Value Analysis, an accounting framework aimed at the quantification of the societal total value of a good, based on input-output modeling, life cycle assessment (LCA), or direct measurement. The framework is rather descriptive and lacks the guidance on how to operationalize the evaluation of necessary data or impacts. Similarly, the Natural Capital Protocol by the Natural Capital Coalition (NCC) also describes a general framework of TCA, the natural capital assessment, but does not provide explicit tools or guidance on the calculation and rather gives ideas for monetization and possible databases. In contrast, the Sustainable Food Trust developed a framework specifically targeted at the true costs of agricultural production per area, with a focus on monetizing the benefits of ecosystem services. Also, the *TEEB Agri-Food Valuation Framework* (TEEB 2015) focuses on agricultural production and defines comprehensively the dependencies and impacts relevant to the externality assessment of an agricultural value chain. Again, it does not provide operationalizable monetization methods. Finally, the System of Environmental-Economic Accounting (SEEA) is a "multipurpose conceptual framework for understanding the interactions between the environment and the economy" but considers only those environmental costs or benefits that are already included in the market.[1] As the previously mentioned approaches lack a detailed description of the needed calculation steps, they are less useful to provide comprehensive insights about the actual calculation of true costs. Therefore, we resort on the approach published by Pieper et al. (2020), which provides these insights and relies on the monetization of impacts resulting from LCA. The authors establish this framework in the case of agricultural production, but it can certainly be approached from all kinds of industries.

Generally, a TCA calculation comprises four steps, each of which will be explained in one section of this chapter. In the following, we will name these steps (1) qualification, (2) quantification, (3) monetization and internalization, and (4) communication. Other approaches may use different nomenclature (e.g., the *TCA Agrifood Handbook* describe them as identification, measuring, evaluation, reporting).

The first step is the *qualification* of the analyzed system boundaries and the *identification* of the therein ecological or social impact categories that are relevant for the evaluated product or process. This is equivalent to the goal and scope definition of LCA (Figure 3.1), a method commonly used for TCA calculations. In this context, environmental impacts are, for example, global

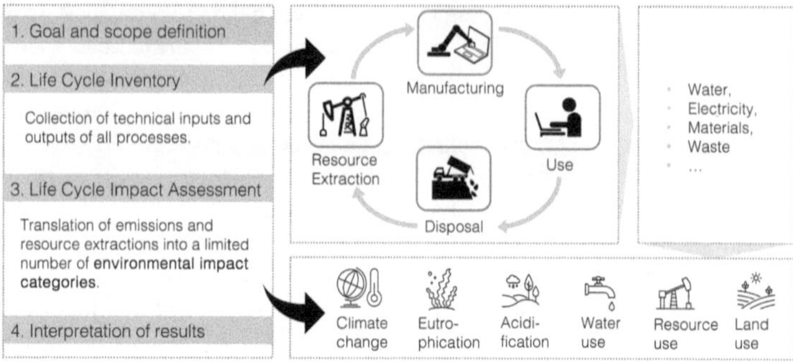

Figure 3.1 Steps of the life cycle assessment (LCA) method as defined in 14040/14044 (ISO 2006a, 2006b)

Figure from Michalke, A., PATOS (www.patos-solutions.de)

warming potential (GWP) induced through production or throughout the process, or water use along the value chain. Social impacts are, for example, fair wages for workers involved in production, or gender equity within the value chain. The step of qualification will be explained in Section 3.2.

The second step comprises the *quantification* of impact that arises along the production processes through the previously qualified impacts. This is likewise done with LCA, which enables the *measuring* of resource use and caused emissions within predefined system boundaries. With Life Cycle Impact Assessment (LCIA) methods, these system models can be translated toward impacts caused within the previously qualified impact categories. While this is an elaborated method for environmental indicators, the social sphere of impacts (assessed through Social LCA or SLCA) is not yet equally well established in LCA literature (Messmann et al. 2020). The step of quantification will be explained in Section 3.3.

The third step is the *valuation* of previously quantified impacts to economic terms via *monetization and internalization* of these impacts. For this, costing methods like damage or abatement costing are used, which underly certain assumptions and uncertainties that must be taken into account when communicating TCA results. With this step, the monetized externalities can also be set into perspective with previous product prices or the financial statements of corporate activities with internalizing them into cost calculations. The step of monetization and internalization will be explained in Section 3.4.

The fourth and final step is the *communication* of TCA results. This must be done in a stakeholder-targeted manner. We will present some ideas for communicating the results of true prices or externality-adjusted business reporting in Section 3.5. Chapters 5 and 6 draw further on this with explaining current TCA application cases and TCA implications for stakeholders, respectively.

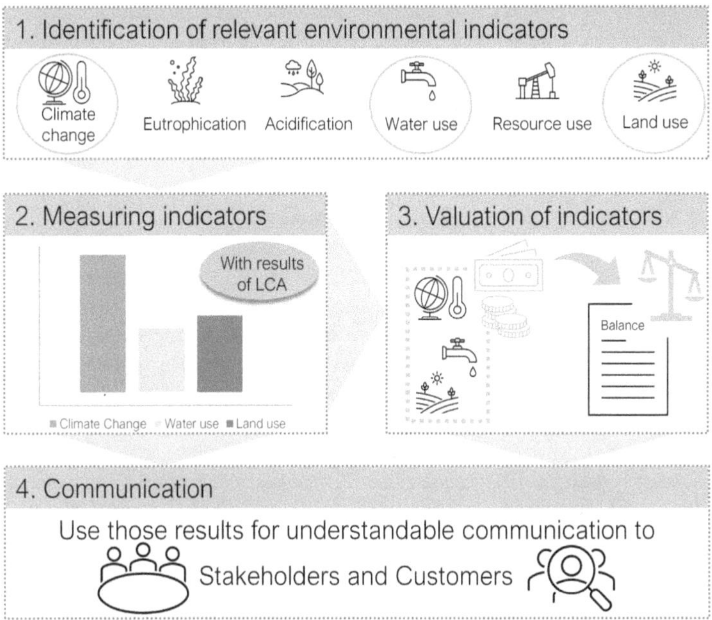

Figure 3.2 Steps of qualifying, measuring, valuing, and communicating externalities

Figure from Michalke A., PATOS (www.patos-solutions.de/true-cost-accounting)

See Figure 3.2 for a visual representation of the method explained herein.

The following chapters describe the four steps of TCA in greater detail and do this with an accompanying exemplary product, i.e., coffee, to make the method more graspable. Coffee production encompasses agricultural processes (as coffee beans must be cultivated), a highly globalized value chain (as they are cultivated in tropical climate, whereas coffee is mostly consumed in the global north), and pressing social issues (as coffee farmers oftentimes are denied sufficient pay for their product). Therefore, a comprehensive example of TCA can be drawn. This is, of course, translatable to other products within other industries. Furthermore, this example of a product-specific TCA can likewise be translated to TCA of a company, when all processes occurring throughout its economic activities are assessed instead of the herein exemplified product-based approach.

3.2 Qualification of relevant indicators

Key issue: What are relevant considerations for qualifying the system that is being analyzed?

To get started with a TCA calculation, one must define the goal and scope of the assessment, meaning the boundaries in which the system to be assessed lies and the use for which the assessment is made. The herein presented method establishes the TCA calculation based on the foundation of LCA. The first substantial part of LCA is the *goal and scope definition*, which must precede all modeling. This requires a deep understanding of the process or product in question and builds the foundation for all further calculation.

A first step is drawing out the *system boundaries*. This requires a definition of all processes that are included in the assessment and ensures an inclusion of all relevant factors. Jacquemin et al. (2012) describe four main boundaries for an LCA: Cradle to gate, gate to gate, cradle to grave, and gate to grave. Cradle to grave describes the full life cycle of a product, for example, whereas cradle to gate describes the system from its very beginning to a certain processing step along the value chain. There could be several "gates" along one value chain of a product. According to Li et al. (2013) this boundary definition describes the constraint criteria of the *economic dimension*, or economic processes considered in the assessment. There is also the *geographical dimension*, which describes from where (i.e., global, continental, regional, national, local, company) data must be collected for the goal of the study. The time frame in which data are collected, and the age of data is determined by the *temporal dimension*. The temporal boundaries are somewhat dependent on the economic dimension: The time frame in which the assessed processes are being executed is the smallest necessary temporal boundary. However, sometimes data are not gathered as primary data at the process itself, but from secondary data from previous assessments or even literature. Therefore, one must decide how "old" data can possibly be to still depict the current process chain under assessment. With a secondary data approach, some questions might limit the time frame in which data are adequate to be used for an assessment of the present situation. A sensible definition of all boundaries is ultimately the responsibility of the company conducting the TCA.

Example:

Economic dimension: The "cradle" for coffee would be the sowing of the coffee seed. Accompanying this are all inputs that are needed for it, like the seeds, fertilizers, water, machinery, etc., with all their inherent processes, like watering, transportation, etc. The gates along the coffee value chain could be the farmgate, when the unprocessed beans leave the farm, the gates of the roastery, or even the gate of a supermarket selling ground coffee for home use. Within the chosen boundaries, all processes with their inherent resource consumption leading up to this specific point are included in the assessment, meaning the gate of "supermarket" will have more processes included – like transportation of the raw material, energy use for storage in the supermarket, etc. – compared to the farmgate. The grave of coffee would be the used coffee grind ending in landfill or used as fertilizer for house plants. This shows that the choice of system boundary exerts a great influence on results of an LCA and consequentially TCA: For example, ending coffee grounds on landfill will cause higher emissions (like methane from biochemical reactions of organic components in landfills; Themelis and Ulloa 2007) along this last step compared to using coffee ground as fertilizer in the home. For our example, we define the boundaries of cradle to farmgate.

Geographical dimension: Since coffee is being produced primarily in countries of the global south (largest producing and exporting are Brazil, Vietnam, and Colombia), however, consumed primarily in the global north (largest importing countries are the United States, Germany, and France; Voora et al. 2019), its value chains span across the whole globe. Therefore, for a cradle to grave approach, for example, data must be acquired globally from the countries tangent to these processes. For our example, we define the boundaries as Brazil, where our specific coffee is farmed.

Temporal dimension: Questions to define this temporal restraint in our example might be: Did the farms producing the coffee beans change their fertilizing regime? Did distributors change from diesel to petrol vehicles? Is the coffee now fairtrade certified to ensure better working conditions at farms? For our example, we pick the boundaries as the last ten years since we would need secondary data for this assessment and decide that these should not be older than a decade.

Another perspective that has to be defined is the *environmental dimension*. The ISO 14044 as the norm of LCA describes that "[the] selection of impact categories shall reflect a comprehensive set of environmental issues related to the product system being studied, taking the goal and scope into consideration" (ISO 2006b). Depending on the later in step (2) used method for impact assessment, LCA provides different impact categories in which the environmental effects are accumulated. Which of these impacts are to be included in the assessment, should be decided following the goal and scope of the TCA. Generally, for a TCA, it is sensible to monetize as many impacts as possible since this brings the calculation closer to the "true cost" of the assessed product or process. Similarly, impact categories for the social assessment should be selected for consideration of the *social dimension* (provided by SLCA; more on this matter in Section 3.3).[2]

Example:

Environmental dimension: Depending on the goal of the study, the impact categories should be chosen based on the needed information for the audience of the particular TCA. Since coffee production is an agricultural process, the impact of land use, for example, is of high interest. Moreover, in this context, the impacts of toxicity (terrestrial, aquatic, or human) from used chemicals, like fertilizers or plant protection, are interesting to assess. In light of current developments regarding the changing climate, global warming potential is always sensible to include. As described, as many impacts as possible should be assessed for TCA. However, for the sake of simplicity of our example, we pick terrestrial ecotoxicity, and global warming potential, as the impacts for assessment.

Social dimension: Similarly, to the environmental dimension, social impact categories should be chosen based on the study's goal. In the example of coffee production, where farming takes place mostly in countries of the global south, factors like living wages or modern slavery play a crucial role. For our example, we pick the social issues of fair wages and forced labor.

It should be noted that the current method of LCA does not enable the qualification (and later quantification) of positive aspects occurring during production or along processes. Merely negative impacts, or damages, can be assessed here. This means that production practices that focus on providing its engulfing system with positive outcomes are not advantaged in LCA over those not doing so. An example would be organic farming practices compared to conventional approaches: Although environmental impacts from organic farming can of course not be avoided completely, positive impacts are likewise generated, like the provision of ecosystem services (e.g., soil carbon sequestration, decreased soil erosion, etc.). Such positive effects are not yet displayed in LCA. Therefore, LCA is often criticized for not depicting the whole environmental effects caused by certain production or processes. However, approaches are being developed to integrate such positive considerations into the assessment methods.

Besides the definition of the aforementioned system boundaries, the *functional unit (FU)* also needs to be chosen. The FU describes the measuring unit based on which the environmental or social impacts are being assessed. This could be, for example, one kg of product, one hour of processing time, or one final product. Depending on the context in which the LCA – and eventually TCA – is executed, the FU must be sensibly chosen. A per kg of product approach, for example, seems senseless if the assessment is done for the production of an electronic device for informational purposes to the customers since, when sold, this product would not be measured per kg but per one entity of product. One must consider as well that the chosen FU has great influence on the results of the TCA. Again, drawing on the example of organic and conventional agriculture, if the production of food is being assessed along the FU of "per hectare of agricultural area", organic agriculture generally creates less environmental impact due to less intensive production practices that are being applied per hectare (like fertilizers or pesticides). If the FU is chosen as "per kg of food product", organic agriculture oftentimes creates more environmental impact, since the product yield per hectare is smaller in most cases, and production practices are being used for less outcome, the created impact therefore is shared by less product units. Therefore, not only the FU must be chosen carefully and targeted toward the specific use case, but results must also be communicated against the background of all system definitions (system boundaries and FU).

Example:

Functional unit: Since our economic boundaries are cradle to farmgate, and coffee is sold by weight, the functional unit for our calculation example is per kg of coffee.

After this thorough qualification of our assessment case, we are able to move on to step (2): Quantification.

3.3 Quantification of environmental and social impact

Key issue: What are the impacts generated throughout the assessed production process?

After the goal and scope are defined, and therefore, the relevant processes and production steps along with all necessary impact categories are clear, these impacts are now being assessed with further steps of LCA. The following explanations largely pertain to environmental aspects of the assessment. The quantification of social impacts is covered later in this section.

The second step of an LCA (after the goal and scope definition) is the *life cycle inventory (LCI)*. The LCI is comprised of an array of different processes going into the production or process. This aims at accumulating all inputs needed and outputs generated for the production in order to subsequently derive the impacts of in- and outputs overall. There are several LCI methods and their different use cases are dependent on the goal and scope, as well as the available time, labor, and money for the assessment (Islam et al. 2016). All inventories partition the assessed product or process into incrementally smaller processes or products to ease the accumulation of all in- and outputs. This presumably becomes clearer looking at Figure 3.3.

Furthermore, *cutoff criteria* are being used to emphasize the more impactful flows of in- and outputs. ISO 14049 explains how a flow can be omitted from further analysis if it contributes less than a predefined percentage of the total mass inflow to the assessed process (ISO 2012). However, Li et al. (2014) argue that this is problematic, since mass alone is not the determining force behind the environmental impact generated by the overall process or products. Therefore, cutoff criteria are useful for a sensible simplification of the inputs and outputs especially when dealing with a product or process that needs a magnitude of upstream processes, and material flows. However, the execution of such criteria can be challenging for the accuracy of the overall assessment.

Regarding *data* necessary for the LCI, existing secondary data can be used for modeling the processes. LCA databases provide data for certain products or processes at different life cycle stages. They are available for various industries, like agriculture (e.g., Agri-footprint, AGRIBALYSE), chemicals (e.g., cm.chemicals), or cross-industry databases (e.g., Ecoinvent, GaBi). Such databases oftentimes draw data from literature and can be seen to provide the

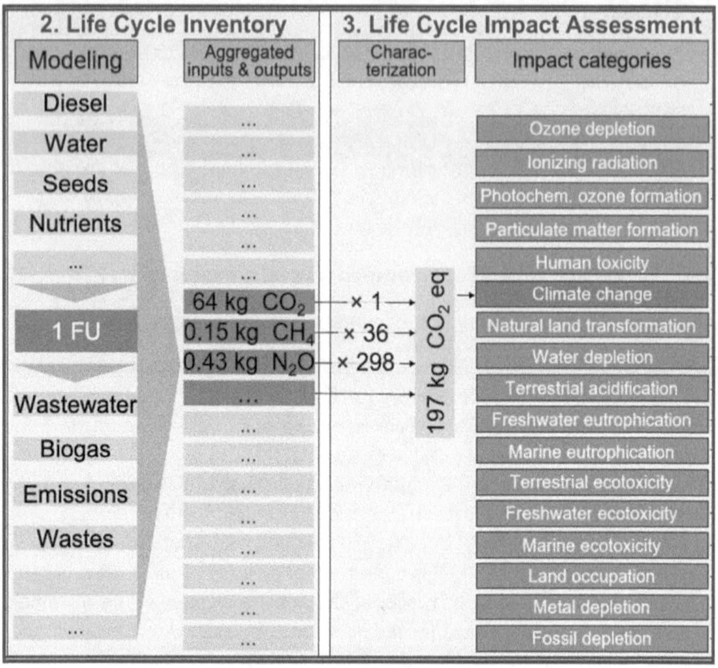

Figure 3.3 Steps 2 and 3 of the LCA (see Figure 3.1), aggregating process steps into inputs and outputs and characterizing those into impact categories

Own figure based on Huijbregts et al. (2016)

"average" data set for a particular product or process that is being assessed within a region or according to certain standards. Each database has different principles of modeling and primary sources, both of which should be understood by the LCA executer to make responsible decisions when choosing the underlying data. If companies want to assess the specific life cycle of their specific product or process, however, they need to gather primary data along the value chain and implement this within their specially applied LCI models. All data, secondary or primary, and modeling are best processed in LCA software tools that facilitate the assessment tremendously.

Example:

Life cycle inventory: Coffee production on farm requires different production steps and processes for the final product at farmgate, like sowing of seeds, using plant protection, harvesting, or sorting.

Cutoff criteria: We will not define a cutoff criterion for this example, as the coffee production process is modeled without one in the picked database.

Data: Agricultural processes like coffee farming are represented in several databases. We will use secondary data modeled in an existing production process provided by the Ecoinvent 3 database and carry out the assessment with the tool SimaPro.

After listing the LCI, and with it all necessary input and generated output (comprised of materials or emissions), follows the *LCIA*. This is the step that translates factual elementary flows to the environmental and social impact categories. The calculation itself is mostly conducted with LCA software, which models the underlying mechanisms as a cause-effect chain from environmental intervention to its resulting impact (Rosenbaum et al. 2017). For this, characterization factors are developed and used to translate the inventory results to the unit of each environmental impact category (ISO 2006b). This makes quantities of substances outflowing from the system comparable to each other. For example, one kg of CO_2 as well as one kg of methane do have influence on the climate, but their impact differs. According to the IPCC (2021) methane is about 30 times as climate damaging as CO_2. Therefore, the characterization factor of Methane is 30-fold that of CO_2 for the impact category of GWP.

Different methods of LCIA exist that each define similar, yet slightly differing impact categories and equally differing characterization factors for them. According to the goal and scope of the assessment, the right LICA method should be picked – and therefore should at least include the impact categories deemed important throughout the qualification process. Examples of such methods are IMPACT2002+ (Humbert et al. 2014), ReCiPe 2016 (Huijbregts et al. 2016), or the Product Environmental Footprint (PEF) (Manfredi et al. 2012). Furthermore, the LCIA method should be compatible with the later on used method for monetization (further explored in Section 3.4). At the end of the LCIA, the overall impact of the product or process is aggregated and presented through the various impact categories. It is possible to aggregate this impact even further to overall damage inflicted from the impact with a weighting of impact toward damage categories, or also called areas of protection. Such damage categories are defined in almost any LCIA method and usually comprise categories along the lines of damages to human health, to ecosystem quality, to climate change, and to resources. Again, the allocation of impact to damage is dependent on the respective LCIA method. This aggregation certainly makes products or processes more comparable to each other (i.e., it is easier to compare between three or four damage categories, rather than 16–20 impact categories), and damages are likely a little more graspable for the common reader of an LCA. However, the aggregation also bears some uncertainties, precision is likely lost along this step, and results consequently are less accurate. The herein presented monetization method works with the monetization of impact categories and therefore weights the different impact categories to one single score.

Example:

Life Cycle Impact Assessment: For our example, we choose the LCIA method ReCiPe (Huijbregts et al. 2016). This method includes the impact categories previously qualified as important for our use (in total, ReCiPe comprises 17 impact categories, which would optimally all be evaluated for a full LCA). It is also compatible with the later used monetization approach, the combination of which is based upon Michalke et al. (2023).

The previously described steps are mostly targeted toward environmental impacts. Regarding the quantification of social impact, the to date developed methods do not seem as cohesive as for environmental aspects. One differentiating factor is that a social assessment based on the FU does not seem sensible: In SLCA, socioeconomic impacts are assessed, which is not linked to physical streams of elementary in- or outputs but rather the geographically specific circumstances a company is situated in as well as the company's dealing with workers, suppliers, and other stakeholders (Wu et al. 2014). Models that calculate impacts for average production chains no longer seem sufficient and data must be collected for explicit products along their value chain. So oftentimes, site-specific data are gathered with surveys or interviews. Approaches for providing contextual information (average country- and industry-specific situation) on social situations within an economy and modeling this along structures of ecological LCAs exist, however, for example, by the Social Hotspots Database (SHDB) (Benoit-Norris and Norris 2015), but authors also caveat how this contextual data must be supplemented or replaced by value-chain-specific data necessary for the evaluation of the assessed process or product. Other approaches are manifold and gathered in several literature reviews (e.g., Wu et al. 2014, Petti et al. 2018).

Example:

Social life cycle assessment: For our example, we chose the assessment within the SHDB, which includes 22 social issues. The impact assessment for this database is called Social Hotspots Index, which weights the risk levels of damaging situations occurring in a country with the working hours needed in this country to generate 1 USD. This yields the average risk hours contributing to the generation of 1 USD for every impact category. Since the SHDB does not list coffee as a specific crop, we use the data set for "other crops, BRA" for extracting the data for forced labor and fair wages in Brazilian Coffee production. This emphasizes the importance of gathering primary data for the specific assessment case in order to best depict TCA for the assessed situation.

Now that all ecological and social damages relevant to the specific system are quantified, they have to be translated into economic terms and compared with current price levels. Therefore, the next step (3) is carried out: Monetization and internalization.

Example results (these are only results for environmental LCA, since we are not further calculating external costs of the SLCA results, see Section 3.4):

Global warming potential: 1 kg of Brazilian coffee produces 15.73 kg of CO_2eq from cradle to farmgate.

Terrestrial ecotoxicity: 1 kg of Brazilian coffee produces 15.09 kg 1.4-DCB eq from cradle to farmgate.

3.4 Monetization and internalization

Key issue: What are the external costs of the assessed product or process and how do they compare with current market prices?

After elementary flows and impacts are quantified, TCA now proceeds with an economic evaluation. The previously yielded impacts are being monetized for twofold aspects: First, this enables a depiction of the "true price" of the assessed product or process; second, this converts all impacts to one unit, making all impact categories, and the assessed products or processes comparable with each other.

There are types of costing methods that go beyond the conventional costing procedure common for an enterprise. Life cycle costing (LCC), for example, includes costs of operation, maintenance, and disposal of a product or process besides the normally accounted acquisition costs (Ciroth et al. 2008, Ciroth 2009), as long as all these costs are borne by at least one actor along the value chain of said product or process (be that producer, processor, user, supplier, etc.) (Swarr et al. 2011). Such costs are in parts tied to ecological or social issues but are no externalities and therefore not body of TCA since they are paid for by some actor along the value chain. Examples are costs of ecotaxes, costs of rehabilitation (after environmental accidents, for example), or cost savings of renewable energy utilization (Kumaran et al. 2001). Nevertheless, with the assessment of such costs, no externality assessment is achieved with LCC as it is approached from a traditional economic lens, which focuses on financial capital, but neglects natural or social capital (Fiedler et al. 2018).

An externality is the loss of welfare – not of a company's profits – due to environmental or social damaging impact (de Bruyn et al. 2018) and is therefore not financial capital and hence externalized from the market. Costing methods translating this externalized impact into monetary terms are mainly approached from two perspectives: While *abatement costs* describe costs of reduction or prevention of environmental impact, *damage costs* describe costs due to damage originating from environmental impact (Amadei et al. 2021). Abatement costs are also called mitigation or prevention cost. This costing approach is oftentimes used to support decision making (political or also internal to companies), because it displays the cost for certain mitigation targets or measures (Huang et al. 2016, Oberpriller et al. 2021). Examples for such measures would be the reduction of air pollution with filters, or using renewable energy sources instead of fossil energy (Guerriero 2020). Damage costs on the other hand are used mostly to measure the effects of inaction or failure to achieve certain sustainability targets (Oberpriller et al. 2021). Regarding the subject of investigation, this can be described as the inaction of policies, producers, or even consumers. Examples for such damage would be the cost of restoring an ecosystem to its original state (e.g., reforestation of previously deforested area) (Ott et al. 2006), or the reimbursement of people suffering from environmental impact (Galgani et al. 2021) (e.g., paying a settlement for contaminating drinking water). Another perspective is measuring the economic impact according to individuals' willingness to pay (WTP). The monetary value for a good with no market price – like environmental impact or integrity – is extracted from human behavior or perception with questions on how valuable a certain service or situation is to the respondents (Bachmann 2011). An example of this approach would be public WTP for reduced environmental issues in an area of their interest, like residents' WTP for less eutrophication of the Baltic Sea (Ahtiainen et al. 2014). WTP – or the money people are willing to give – can be transferred to both abatement and damage costs, depending on the subject of questioning (i.e., "How much are you willing to pay for this mitigation measure?" or "How much are you willing to pay for this restoration measure?").

To economically evaluate LCA results, methods use a mix of these costing types to assess the previously quantified impacts – which have differing category-specific units – monetarily. Arendt et al. (2020) compare nine different monetization methods of LCA, giving a quite comprehensive overview of existing approaches. They find that damage costs are used by most approaches; however, oftentimes, within one monetization framework, costing methods are mixed. For example, the impact category of global warming is mostly evaluated with abatement costs. It also becomes clear that costing approaches yield vastly different results: For certain impact categories, some approaches find costing factors that deviate up to eightfold compared to the mean of all approaches. Amadei et al. (2021), investigating similar approaches as Arendt et al. (2020), also find that some impact categories (like climate change) are analyzed thoroughly in literature, while others (like eutrophication) lack scientific attention. All this shows the importance for a differentiated communication of results and uncertainties of TCA (more in Section 3.5).

To monetize the specific impact categories previously qualified as important to the process or product assessed, one needs to choose a costing approach that is compatible with the LCIA method that was being used for quantifying the environmental or social impact. The weighting or monetizing factors depend on the characterization models used for impact assessment (de Bruyn et al. 2018).

Example:

Monetization of impact: For our example, we choose monetization methods compatible with our LCIA method ReCiPe. This is, for example, the *Environmental Prices Handbook* (EPH) (de Vries et al. 2024). This is based on the approach taken by Michalke et al. (2023) and updated. We picked global warming and ecotoxicity as impacts to be assessed. Both their monetization methods based on de Vries et al. (2024) will be explained exemplarily in the following:

Global warming: Global warming is an impact category on which international agreements for policies have been reached, which is why abatement costing would be a sensible approach. The pricing of CO_2eq (the unit in which global warming is presented in ReCiPe) is based on price trends from the European Union Emission Trading System (ETS) and extended to present an economy-wide CO_2 price. Depending on the time frame in which the herein modeled mitigation targets are observed, the monetization factor will vary: The further in the future the target is, the higher the costs per CO_2eq, since it will be harder, and therefore more costly, to control climate change along a progressing timeline. According to these models, EPH arrives at a costing factor of 0.13€per kg of CO_2 emission (about 130€per ton). This seems rather low when comparing this with other evaluations of the cost of GWP: The German Federal Environmental Agency (UBA) declares this at 200€per ton, Ricke et al. (2018) indicate this at 370€, and the True Pricing Foundation establishes 150€per ton of CO_2. This shows the differences also in costing scenarios that will certainly impact the results of a TCA.

Terrestrial ecotoxicity: Ecotoxicity is the impact of toxic substances on nonhuman organisms, which can in excess pollute water basins and soils, and cause accumulation in food chains, for example. This can lead to changes in the circumstances and conditions of exposed invertebrates and vertebrates, for example. In EPH, monetization of ecotoxicity is approached with damage costing, and in particular with its damages on biodiversity. EPH describes a cost factor for terrestrial ecotoxicity at €0.00095 per kg of 1,4 Dichlorobenzene (DB) (the unit in which ecotoxicity is measured; 1.4 DB is a molecule commonly used as disinfectant, pesticide, or deodorant, for example). EPH also concede that environmental prices for ecotoxicity involve great uncertainties, which we want to caveat here as well.

Monetized example: With these cost factors and the results from the environmental LCA, the costs of global warming for 1 kg Brazilian coffee from cradle to farmgate are (15.73 kg CO_2 eq * 0.13€ per kg CO_2 eq =) 2.04€. Likewise, the costs of terrestrial ecotoxicity are (15.09 kg 1.4-DCBeq * 0.00095€ per kg 1.4-DCBeq =) 0.01€. In total, this would be a price surcharge (for only two environmental indicators) of 2.05€ per kg of Brazilian coffee.

The monetization of social factors has not yet been attempted based on a conventional LCA approach. The results from 3.3, yielded from the SHDB, are measured in average risk hours per generated USD. But morally, how can a price be attached to this unit? What is a safe working environment worth? Other approaches, like the Oiconomy system, provide a framework that works along the principles of LCA (as data are gathered along a certain life cycle, impact categories are measured with the quantification of indicators). However, here, every supply chain actor is responsible to measure their hidden social costs themselves and to include this in their financial statements (Croes and Vermeulen 2015). This is more in line with the primary data approach and with some examples already established in practice (like the fairtrade-label, or the company *Truesday*, both explored in Chapter 5). We will not attempt to monetize social impact within our example at this point, since methods for SLCA monetization are not yet explored sufficiently and we would not do this sensitive topic justice at this point in time.

Now that relevant external costs are monetized, these can be put into perspective with the current market price or cost calculation of the assessed product or process and are therefore internalized. This can be discussed twofold: When added to the current pricing, the monetized externalities can reveal the "true costs" of the assessed product or process (without claim to completeness due to previously mentioned uncertainties to the method, and limitations to the defined system). If put into relation to the current prices, the monetized externalities depict price surcharges that describe current market distortions; it also makes different products or processes comparable in terms of the inadequacy of their currently underlying pricing mechanisms. Data for these market prices are of course incumbent on the department of the company responsible for pricing policy and strategy.

Example:

Internalization of externalities: For our example, we use data for current producer prices of coffee for the country of Brazil based on *FAOSTAT* (Food and Agriculture Organization Statistics). This was, in 2022, 3162.7 USD per ton, which translates to roughly 3.01€ per kg of coffee. Including the surcharge of 2.05€per kg of Brazilian coffee, the true costs (or "more true", considering only two environmental impacts are measured) amount to about 5.06€ per kg of coffee, which is a surcharge of more than 68% of the producer price.

3.5 Communication

Key issue: How can TCA results be communicated successfully?

Results from the monetized social or environmental impact in comparison with current market prices can be used for different ways of communication. For this step, it is crucial to know the audience of the TCA and to design the communication along their understanding and personal aims.

TCA can be used as a communication tool itself. Of course, currently the depiction of "true prices" or the conduction of a TCA for controlling purposes are not yet mandatory for businesses. Results of TCA, however, can be used for transparency communication, nonetheless. Oftentimes, environmental or social assessments come along with overwhelmingly detailed information on different impacts across different stages of the process or production. With a monetization of aspects from different backgrounds, these different impacts are unified and can be compared against each other, or hotspots along the system's boundaries become visible. This supports decision making both internally and externally: Processes of the company can be improved according to the points within the system that show the highest externalities overall. Therefore, lowering externalities for the company can be purposefully targeted internally. Furthermore, companies can use TCA results to communicate their compliance with sustainability standards, for example, to monitoring entities, or as proof of their inherent sustainability endeavors for customers.

To optimally enable this, the communication of the results needs to be designed and addressed in a recipient-specific manner. If the company strives to improve processes, the results must be prepared in such a way that enables internal decision-makers to identify the processes that are responsible for high environmental or social externalities; an aggregated result to one single number of external costs might not be sufficient for this. Contrary, for communicating to customers, for example, a more streamlined result should be communicated. Customers do not necessarily need to know which process exactly drives externalities related to ecotoxicity, for example, if they want to make a sustainability-informed purchasing decision in favor of or against a certain product. The overall result would suffice here, as it discloses information on the company's or the product's sustainability performance in an easily understandable way.

In Chapter 5, we explore cases that already implement TCA in business activities and with it different approaches of communicating its results. Furthermore, Chapter 6 ends this book with a look at the impact and targeting of TCA for all relevant stakeholder groups.

Notes

1 The previously presented list of TCA methods is adapted from Fiedler et al. (2018).

2 Caution should be taken when communicating results of this assessment, since many structures along a globalized value chain have developed through historically grown power imbalances and are perpetuated and upheld through a globalized market structure. Placing blame on certain players of the value chain (e.g., coffee farmer who might be unable to pay living wages to their workers due to highly competitive market scenarios) is difficult. However, qualifying such social hotspots along the value chain can be very valuable for targeted people.

References

Ahtiainen, H., Artell, J., Czajkowski, M., Hasler, B., Hasselström, L., Huhtala, A., Meyerhoff, J., Smart, J. C. R., Söderqvist, T., Alemu, M. H., Angeli, D., Dahlbo, K., Fleming-Lehtinen, V., Hyytiäinen, K., Karlõševa, A., Khaleeva, Y., Maar, M., Martinsen, L., Nõmmann, T., … Semeniene, D. (2014). Benefits of meeting nutrient reduction targets for the Baltic Sea – A contingent valuation study in the nine coastal states. *Journal of Environmental Economics and Policy*, *3*(3), 278–305. https://doi.org/10.1080/21606544.2014.901923

Amadei, A. M., Laurentiis, V. D., & Sala, S. (2021). A review of monetary valuation in life cycle assessment: State of the art and future needs. *Journal of Cleaner Production*, *329*, 129668. https://doi.org/10.1016/j.jclepro.2021.129668

Arendt, R., Bachmann, T. M., Motoshita, M., Bach, V., & Finkbeiner, M. (2020). Comparison of different monetization methods in LCA: A review. *Sustainability*, *12*(24), 10493. https://doi.org/10.3390/su122410493

Bachmann, T. M. (2011). Optimal Pollution: The Welfare Economic Approach to Correct Market Failures. In: Nriagu, J. O. (eds) *Encyclopedia of Environmental Health* (S. 264–274). Elsevier. https://doi.org/10.1016/b978-0-444-52272-6.00301-9

Benoit-Norris, C., Norris, G. A. (2015). Chapter 8: The Social Hotspots Database Context of the SHDB. In *The Sustainability Practitioner's Guide to Social Analysis and Assessment* (52–73).

Ciroth, A. (2009). Cost data quality considerations for eco-efficiency measures. *Ecological Economics*, *68*(6), 1583–1590. https://doi.org/10.1016/j.ecolecon.2008.08.005

Ciroth, Thomas, A., Huppes, G., Klöpffer, W., Rüdenauer, I., Steen, B., & Swarr. (2008). *Environmental Life Cycle Costing* (Hunkeler, D., Lichtenvort, K., & Rebitzer, G. (eds)). CRC Press.

Croes, P. R., & Vermeulen, W. J. V. (2015). Comprehensive life cycle assessment by transferring of preventative costs in the supply chain of products. A first draft of the Oiconomy system. *Journal of Cleaner Production*, *102*, 177–187. https://doi.org/10.1016/j.jclepro.2015.04.040

de Bruyn, S., Bijleveld, M., de Graaff, L., Schep, E., Schroten, A., & Vergeer, R. (2018). *Environmental Prices Handbook*. Available at: https://cedelft.eu/wp-content/uploads/sites/2/2021/04/CE_Delft_7N54_Environmental_Prices_Handbook_EU28_version_Def_VS2020.pdf

de Vries, J., de Bruyn, S., Boerdijk, S., Juijn, D., Bijleveld, M., van der Giesen, C., Korteland, M., Odenhoven, N., van Santen, W., & Pápai, S. (2024). *Environmental Prices Handbook 2024: EU27 version*. Available at: https://cedelft.eu/wp-content/uploads/sites/2/2024/12/CE_Delft_230107_Environmental-Prices-Handbook-2024-EU-version_def.pdf

Ernst & Young (EY). (2016). *Total value impact valuation to support decision-making*. Available at: https://tca2f.org/wp-content/uploads/2019/09/ey-total-value-impact-valuation-to-support-decision-making.pdf

Fiedler, K., Lord, S., & Czarnezki, J. (2018). Life cycle costing and food systems: Concepts, trends, and challenges of impact valuation. *Michigan Journal of Environmental & Administrative Law*, *8*(1), 1. https://doi.org/10.36640/mjeal.8.1.life

Guerriero, C. (2020). *Cost-Benefit Analysis of Environmental Health Interventions* (S. 111–127). Academic Press.

Huang, S. K., Kuo, L., & Chou, K.-L. (2016). The applicability of marginal abatement cost approach: A comprehensive review. *Journal of Cleaner Production*, *127*, 59–71. https://doi.org/10.1016/j.jclepro.2016.04.013

Huijbregts, M. A. J., van Z, R. R., Steinmann, Z. J. N., Elshout, P. M. F., Stam, G., Verones, F., Vieira, M. D. M., Hollander, A., & Zijp, M. (2016). *ReCiPe 2016- A harmonized Life Cycle Impact Assessment method at midpoint and endpoint level Report I: Characterization*. Available at: https://rivm.openrepository.com/bitstream/handle/10029/620793/2016-0104.pdf?sequence=3&isAllowed=y

Humbert, S., Jolliet, O., Margni, M., Charles, R., Payet, J., Rebitzer, G., & Rosenbaum, R. K. (2014). *A user guide for the Life Cycle Impact Assessment Methodology*

IMPACT 2002+. Available at: https://quantis.com/pdf/IMPACT2002_UserGuide_for_vQ2.21_30April2014a.pdf

International Standards Organization. (2006a). *ISO 14040- Environmental management—Life cycle assessment—Principles and framework*. Available at: https://www.iso.org/standard/37456.html

International Standards Organization. (2006b). *ISO 14044:Environmental management—Life cycle assessment—Requirements and guidelines*. Available at: https://www.iso.org/standard/38498.html

International Standards Organization. (2012). *ISO ISO/TR 14049:2012-06 Umweltmanagement—Ökobilanz—Beispiele zur Anwendung von ISO 14044 zur Festlegung des Ziels und des Untersuchungsrahmens sowie zur Sachbilanz*. Available at: https://www.iso.org/standard/57110.html

IPCC. (2021). *Climate Change 2021: The Physical Science Basis. Contribution of Working Group I to the Sixth Assessment Report of the Intergovernmental Panel on Climate Change*. Cambridge University Press. Available at: https://doi.org/10.1017/9781009157896

Islam, S., Ponnambalam, S. G., & Lam, H. L. (2016). Review on life cycle inventory: Methods, examples and applications. *Journal of Cleaner Production*, *136*, 266–278. https://doi.org/10.1016/j.jclepro.2016.05.144

Jacquemin, L., Pontalier, P.-Y., & Sablayrolles, C. (2012). Life cycle assessment (LCA) applied to the process industry: A review. *The International Journal of Life Cycle Assessment*, *17*(8), 1028–1041. https://doi.org/10.1007/s11367-012-0432-9

Kumaran, D. S., Ong, S. K., Tan, R. B. H., & Nee, A. Y. C. (2001). Environmental life cycle cost analysis of products. *Environmental Management and Health*, *12*(3), 260–276. https://doi.org/10.1108/09566160110392335

Li, T., Zhang, H., Liu, Z., Ke, Q., & Alting, L. (2013). A system boundary identification method for life cycle assessment. *The International Journal of Life Cycle Assessment*, *19*(3), 646–660. https://doi.org/10.1007/s11367-013-0654-5

Manfredi, S., Allacker, K., Pelletier, N., Chomkhamsri, K., & de Souza, D. M. (2012). *Product environmental footprint (PEF) guide*. Available at: https://ec.europa.eu/environment/archives/eussd/pdf/footprint/PEF%20methodology%20final%20draft.pdf

Messmann, L., Zender, V., Thorenz, A., & Tuma, A. (2020). How to quantify social impacts in strategic supply chain optimization: State of the art. *Journal of Cleaner Production*, *257*, 120459. https://doi.org/10.1016/j.jclepro.2020.120459

Ott, W., Baur, M., Kaufmann, Y., Frischknecht, R., & Steiner. R. (2006). *Assessment of biodiversity losses—Monetary valuation of biodiversity losses due to land use changes and airborne emissions*. Deliverable D.4.2.-RS 1b/WP4 of NEEDS – New Energy Externalities Developments for Sustainability.

Petti, L., Serreli, M., & Cesare, S. D. (2018). Systematic literature review in social life cycle assessment. *The International Journal of Life Cycle Assessment*, *23*(3), 422–431. https://doi.org/10.1007/s11367-016-1135-4

Pieper, M., Michalke, A., & Gaugler, T. (2020). Calculation of external climate costs for food highlights inadequate pricing of animal products. *Nature Communications*, *11*(1). https://doi.org/10.1038/s41467-020-19474-6

Ricke, K., Drouet, L., Caldeira, K., & Tavoni, M. (2018). Country-level social cost of carbon. *Nature Climate Change*, *8*(10), 895–900. https://doi.org/10.1038/s41558-018-0282-y

Rosenbaum, R. K., Hauschild, M. Z., Boulay, A.-M., Fantke, P., Laurent, A., Núñez, M., & Vieira, M. (2017). Life Cycle Impact Assessment. In *Life Cycle Assessment* (S. 167–270). Springer. https://doi.org/10.1007/978-3-319-56475-3_10

Swarr, T. E., Hunkeler, D., Klöpffer, W., Pesonen, H.-L., Ciroth, A., Brent, A. C., & Pagan, R. (2011). Environmental life-cycle costing: A code of practice. *The International Journal of Life Cycle Assessment, 16*(5), 389–391. https://doi.org/10.1007/s11367-011-0287-5

TEEB. (2015). *TEEB for agriculture & food: An interim report.* United Nations Environment Programme. Available at: https://www.teebweb.org/wp-content/uploads/2015/12/TEEBAgFood_Interim_Report_2015_web.pdf

Themelis, N. J., & Ulloa, P. A. (2007). Methane generation in landfills. *Renewable Energy, 32*(7), 1243–1257. https://doi.org/10.1016/j.renene.2006.04.020

Voora, V., Bermudez, S., Larrea, C., & Baliño, S. (2019). *Global market report: Coffee (sustainable commodities marketplace series 2019).* International Institute for Sustainable Development, State of Sustainability Initiatives. Available at: https://www.iisd.org/system/files/publications/ssi-global-market-report-coffee.pdf

Wu, R., Yang, D., & Chen, J. (2014). Social life cycle assessment revisited. *Sustainability, 6*(7), 4200–4226. https://doi.org/10.3390/su6074200

Part III

True Cost Accounting in Business Practice

4 True Cost Accounting as Part of Accounting

Abstract

The previous chapters were dedicated to the discussion of relevance, provenience, and realization of True Cost Accounting (TCA). The present chapter aims to illuminate the embedding of TCA in the corporate context. Here, we focus on several dimensions. First, we discuss, why TCA constitutes a natural expansion of classical accounting. In this context, we touch on the two subareas of accounting, financial and management accounting, and describe how TCA can add to the different accounting functions. In the second section of this chapter, we concentrate on the major current developments in accounting related to the most important stakeholders of a company, i.e., society, financiers, suppliers, and customers, and discuss the relevance of TCA for these developments. The third section is dedicated to discussing the relationship of TCA to other sustainable information systems. It aims to identify the links between TCA and these other tools, thereby outlining a comprehensive information and early detection system. The final section provides an overview of possible expansion stages when implementing TCA to help companies to get started with TCA.

Key takeaways

After reading this chapter, you will:

- Understand True Cost Accounting (TCA) as a natural progression of classical accounting
- Know the latest trends in accounting for which TCA can provide information
- Understand how to integrate TCA with other sustainable information tools
- And know the reasonable expansion stages of a TCA implementation

4.1 TCA as enrichment of "classical" accounting

Key issue: Why is TCA a natural progression of classical accounting?

Accounting has a long history and can be traced back to the Mesopotamian civilization (Giroux, 2017). Its development is closely intertwined with the

DOI: 10.4324/9781003345190-7

emergence of complex social and economic structures. These structures required the coordination of various economic activities, such as growing, storing, and distributing food for larger populations, which, in turn, needed a comprehensive quantification and documentation. Until today these two tasks constitute accounting's core function in the context of corporate activities to allow for managerial decisions and their monitoring by third parties. In this sense, accounting can be broadly defined as the quantification and documentation of resource allocation, output generation, and of monetary flows in companies. To reach this aim several techniques, like cost accounting approaches or budgeting methods, have been developed and to cope with the complexity of modern companies accounting has separated in two broad streams, i.e., financial and management accounting (FA and MA; Infobox 4.1). While MA

Infobox 4.1: Financial and Management Accounting

Financial accounting is dedicated to the gathering of mainly financial information of a company and its activities to condense it to statements, which provide insights regarding a company's financial performance. Examples are the income statement, the balance sheet, and the cash flow statement. The two mentioned first are based "on accrual accounting information which is designed to reflect the financial impact of transactions, transformations or external events on the assets, liabilities and equity of a company" (Schaltegger and Burritt 2010, p. 376). The balance sheet indicates the value of fixed and current assets, equity, and liabilities at a particular point in time. By comparing the balance sheets over several time periods, one can identify changes that indicate value creation or value distortion in an economic sense. It is supplemented by the income statement, which provides information about monetized resource consumption and creation of goods within a period. It is period-related, i.e., it provides information illustrating the reasons for changes within the value of assets, inventories, liabilities, etc. depicted in the balance sheet. For this purpose, deferrals must be made. For example, a machine purchased in one period is not fully recognized in that period but depreciated over its lifetime. The cash flow statement provides information about the cash inflows and outflows over a period and indicates the state of liquidity, i.e., the ability of a company to meet payment obligations at any time. The compilation of these statements is highly regulated by standard setters and legal requirements. Moreover, in different cultural areas different rules apply. Although there are efforts to align the various standards, differences are still found. In contrast, as stressed by Burritt (2005, p. 24), management accounting "has a focus on internal decision-making about business, it looks at past information to guide future decisions, cost receives the greatest attention, manufacturing is the value chain function most widely considered, and the emphasis is control". In this sense, management accounting is dedicated to the provision of partly financial and increasingly nonfinancial information to decision-makers related to strategy formulation, allocation of physical resources, cost planning and control, and performance measurement and evaluation (Burritt 2005). This information is intended to provide insights into operational processes that are unadulterated by selection rights, i.e., different possibilities to account for certain business activities, which are common in financial accounting. In addition, this area of accounting is aimed at activities related to the main purpose of the business model. Cost accounting is an important part of management accounting. One major aspect of it is to allocate overhead costs as fairly as possible to the source of the costs to allow for reasonable cost allocation, cost management, and price setting. Thus, TCA is particularly related to this part of accounting, as it also aims to assess resource consumption and monetizes its impacts, but it has a broader perspective also entailing externalities, which are neglected by cost accounting thus far because they do not lead to any financial impact on the company.

aims at internal addressees, particularly decision-makers, financial accounting provides information of the corporate activities to external addressees, like investors, governmental institutions, and the broader public. This difference regarding addressees is also the reason for the different treatment by legal regulation: While FA is subject to statutory regulation to protect the addressees, MA is free in its design, i.e., companies can frame it according to their needs. Moreover, the former is dedicated to the compilation, preparation, and publication of financial and, increasingly, nonfinancial information on corporate performance, whereas the latter covers activities related to cost accounting, planning, and budgeting as well as performance measurement. Although these explanations suggest a very strong separation of the two accounting strands, in practice they are connected with one another. The following definition of MA by the IMA (2019, p. 4) illustrates this:

> Management accounting is a profession encompassing finance and accounting professionals working inside organizations. Their role involves partnering in management decision making, devising planning and performance management systems, and providing expertise in financial reporting and control to formulate and implement an organization's strategy.

Despite their long tradition and highly differentiated design, especially in larger companies, both accounting domains are facing growing criticism over many years (Schaltegger and Burritt 2010). With respect to *FA* Schaltegger and Burritt (2010) identify four major areas of critique: (1) The first critical aspect comprises the narrow legal perspective on the relevant corporate boundaries, i.e., companies are accountable if they directly harm customers or investors, but they do not have to fear consequences if they contribute to the reduction of biodiversity or if their products contribute to ocean litter. This makes companies responsible for only part of the consequences of their actions. Therefore, FA, as it is designed so far, is not able to depict the actual impact a company has on its environment (Bebbington and Larrinaga 2014), because the boundaries of the considered corporate area of influence are framed in accordance with the cash flows the company receives or has to pay to different contractual partners (suppliers, employees, shareholders). And these cash flows do not fully reflect all resource consumption. (2) Accounting is closely linked to the primacy of profit maximization. Thus, it is deeply embedded in this particular perspective on the social sphere and therefore unable to properly appreciate and evaluate the pursuit of other goals than gaining monetary profits. (3) FA methods are not designed in a way that allows for reasonable evaluation of the impacts from corporate activities on the environment, i.e., it lacks the appropriate tools to perform the needed calculations. (4) Finally, the focus on monetary values in FA impedes a comprehensive assessment beyond the financial dimension because a considerable part of a company's impacts on the ecological and social sphere is difficult to monetize. As highlighted by Schaltegger and Burritt (2010, p. 376) *MA* faces criticism as well, comprising aspects like "the arbitrary use of cost allocation, the dominance of financial accounting rules, a narrow focus on manufacturing costs, and a focus on short term decisions rather than strategic decisions". Thus, MA entails similar shortcomings as FA, which is hardly surprising, since both are based on the same foundation and were developed under the same assumptions (primacy of profit maximization). To cope with these shortcomings social and environmental accounting was developed (Infobox 4.2).

However, these accounting approaches are complicated and have to face a major challenge, namely "the need to address the entity concept and focus on eco-systems and their carrying capacities, thresholds and cumulative effects" (Schaltegger and Burritt 2010, p. 377). Thus, scholars have casted doubts of whether accounting actually can be further developed so that it can cope with all these challenges (Hines 1991, Milne 1996, McKernan 2007, Schaltegger and Burritt 2010). As the discussion in Section 4.2 will show, this debate has been caught up with reality as regulators have decided to increasingly demand social and environmental reporting, investors get under pressure to judge their investments against the background of sustainability, and supply chains are required to be managed in a more sustainable way (e.g., European Parliament

Infobox 4.2: Social, Environmental, and Sustainability Accounting

Accounting related to sustainability issues has a checkered history, with the social perspective dominating at times and the ecological perspective at others (Bebbington 2021). The current debate about sustainability accounting has its roots in the discussion about more accountability of companies conducted in the 1960s. These debates led to the development of social accounting, which gained momentum in the 1970s (Gray 2002). In the 1980s the focus changed from social to environmental issues (Owen 2008), due to increasing public pressure resulting from scandals, misconduct, and litigation (Bebbington and Larrinaga 2014). According to Mathews (1997), in the period of 1990–1995 environmental issues strongly dominated over social issues. Due to scandals, for example, in the textile industry, in combination with the pressure exerted by ecological crises, today both social and environmental aspects are equally important, resulting in the development of combined perspectives, which are partly discussed under the umbrella term of sustainability accounting. However, as indicated by Schaltegger and Burritt (2010), the term sustainability accounting is somewhat blurry, and it mixes up with social, ecological, or ethical accounting. A definition by Gray (2010, p. 12) illustrates this: "Social accounting as I use the term here should be taken to embrace social, environmental, ethical, responsibility and sustainability accounting, accountability, reporting, auditing, investment, and costing and management". Additionally, according to Zvezdov and Schaltegger (2013, p. 2363)

> [s]ustainability accounting entails systems, methods, and processes of creating sustainability information for transparency, accountability, and decision-making purposes. This includes the identification of relevant sustainability issues of the company, the definition of indicators and measures, data collection, overall performance tracking and measurement, as well as the communication with [...] internal and external information recipients.

Zvezdov and Schaltegger (2013) categorize these components in three domains: (1) Sustainability information generation and management, (2) sustainability management control, and (3) sustainability reporting. The first domain deals with questions like "Who plays a role in sustainability accounting?", "Which kind of information is needed?", and "Why is it needed or by whom is it used?". The second domain encompasses a broad set of activities, which are designed to achieve the organizational sustainability goals. The third domain deals with the publication and communication of corporate sustainability activities to external parties. TCA can be related to all three areas. Its implementation requires the clarification of areas of responsibility to enable the generation of the individual pieces of information and a structuring of the related processes. The implementation of TCA is always associated with an objective, e.g., improve sustainable production processes, provide information to customers, or align sustainable processes with suppliers, which must be clearly articulated during implementation. Finally, TCA provides relevant information that can be included in external reporting.

and Council 2013, 2014, 2019, 2020, 2022). Accounting must therefore respond and evolve accordingly. Also, from an academic perspective, the pressure is high, as Bebbington and Thomson (2013, p. 282) stress: "Accounting, however, is a crucial technology that might support sustainable development transitions as long as it infuses the rationales and concerns of this more broadly framed debate". Thus, accounting has to provide answers regarding the measurement, analysis, evaluation, communication, and management of particular sustainability issues to which companies are exposed (Schaltegger and Burritt 2010).

The implementation of True Cost Accounting (TCA) is one way to initiate further development and integrate it in a structured manner in companies. Due to its cost accounting focus, this system is compatible with corporate management control systems. In detail, TCA can provide a value added to the following functions of FA and/or MA:

- *Quantification (MA and FA):* To make corporate processes comparable accounting has to use a unifying scale. This is achieved by translating all relevant business transactions into monetary values. Accordingly, transactions at the real economic level, e.g., the consumption of resources for the production of goods, and their results, e.g., changes in inventories, are valued in monetary terms and summarized in a set of accounts, e.g., the income statement, cost accounting, or the balance sheet. TCA provides a methodology to enrich this quantification process by quantifying impact from resource consumption, which has so far not yet been evaluated. This also comprises issues, which are beyond the capabilities of classical accountants, like the contribution of corporate activities to CO_2 emissions to the atmosphere or the accumulation of nitrate in soils and groundwater.
- *Documentation (MA and FA):* The documentation function consists of writing down the evaluated and quantified corporate processes in a

rule-based manner. This process either is dictated by external regulation or it is designed by management accountants to reasonably support managerial decision making. As information generated by TCA also is codified in monetary terms, it can be easily integrated into this documentation and thereby enrich the database. By performing TCA on a regular basis, companies can build a trajectory over several years, which provides insights into the success of efforts devoted to the achievement of goals along ecological and social dimensions.

- *Planning and control (MA):* Planning comprises the forecasting of resource needs and efficient resource allocation. It also comprises risk forecasting and risk management. To inform management decisions for the future, planning always has to be accompanied by control activities, which identify and analyze deviations between the planned and the realized outcomes of corporate activities. As an accounting tool with close relations to cost accounting, TCA can be integrated into these planning and control cycles and as previously mentioned it offers a quantitative basis to control and improve the achievement of ecological and environmental goals. Moreover, as TCA aims to incorporate externalities, which are overlooked by the classical accounting perspective, it can also provide input into risk forecasting: It sheds more light on possible risky resource consumption, like the cooperation with suppliers that use socially and ecologically questionable methods.
- *Publication (FA):* FA provides a wide range of information to investors to supplement their investment decisions, to the state for the purpose of taxation, and to the interested public. As previously mentioned, and further discussed in Section 4.2, regulation is changing considerably in the European Union and a debate about possible changes is also going on in other parts of the globe. New regulations in the European Union oblige a growing number of companies to report on environmental and social aspects in the sense of double materiality. As discussed in Chapter 1, this means that they must provide information on the social and environmental aspects that influence their business activities, as well as the extent to which their business processes impact these areas. So far, FA lacks a comprehensive tool set to generate and edit this information. TCA is dedicated to generating parts of this information and thereby can enrich FA.

Thus, TCA can be seen as a natural progression of classical accounting. However, it also contains the need for a recalibration of the corporate perspective. How far this recalibration affects corporate processes depends on the level of integration of TCA within these processes. Yet, regardless of the degree of integration, even the first steps toward a functional TCA require a willingness to consider the influence of corporate activities on environmental and social aspects as relevant, even if these influences are not associated with direct costs for the company. This perspective is not compatible with a pure shareholder value orientation.

4.2 Current trends in accounting

Key issue: In which areas can TCA make a difference by providing relevant information?

Accounting is a constantly evolving field, which has had to cope with various changes in the past, like the challenges of increasingly globalized acting companies and the digitization of processes. Economic crises like the oil price shock in the 1970s, the financial crisis 2007/2008, or social crises such as the corona pandemic also challenged the field. As the literature shows, these crises always had an important impact on the development of accounting (e.g., Endenich 2014). Also, the focus on social and ecological accounting is not a new development of recent years but indeed has a long tradition (Infobox 4.2). TCA builds on this tradition and provides important impulses for several developments in the context of sustainability-oriented corporate management, which are related to the main stakeholders *society*, *financiers*, *suppliers,* and *customers*. Figure 4.1 indicates the bidirectional relations between the focal company and these stakeholders.

Society. Standard setters and governmental institutions have started to change regulation all over the world. Certain information related to environmental and social aspects affecting the financial performance of a company had to be disclosed since the beginnings of reporting (Tschopp et al. 2012). However, in the last decades, the discussion around these issues became much more intense (Tschopp et al. 2012). In 2001, the European Union started a process of increasing regulation with the dissemination of the Green Paper with the title Promoting a European Framework for Corporate Social Responsibility. This was intended to start a debate around social and ecological issues in the corporate context (European Commission 2001). This discussion was

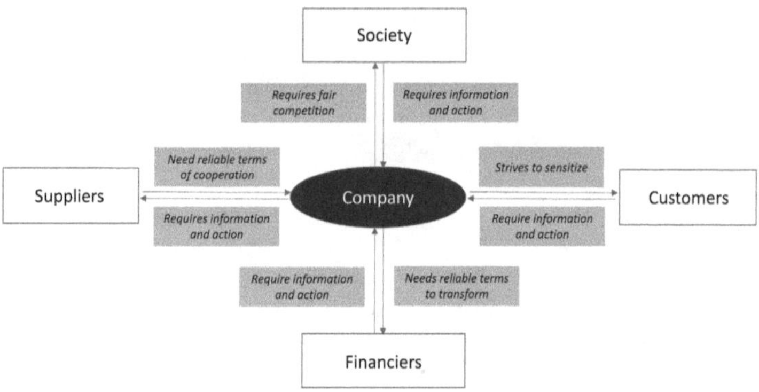

Figure 4.1 Bidirectional relations between the focal company of accounting and surrounding stakeholders

further fueled by the call of the European Commission in 2006 to make the European Union a pioneer in the field of CSR (European Commission 2006). In 2011, the European Union published its strategy to foster CSR between 2011 and 2014 to become a leader in this area (European Commission 2011). Two years later it disseminated the guideline 2013/34/EU for sustainable reporting (European Parliament and Council 2013). This guideline was supplemented by the guideline 2014/95/EU (European Parliament and Council 2014). This, in turn, will be further developed in the Corporate Sustainability Reporting Directive (CSRD) to address existing criticisms (European Commission 2022). As a result of the transposition of these directives into national law, sustainable reporting is becoming mandatory for a growing number of companies. Further, the Sustainability Finance Disclosure Regulation (SFDR) (2019/2088) requires companies in the financial sector to inform about the degree of sustainability of their investment activities (European Parliament and Council 2019). Finally, the European Union developed a taxonomy related to sustainable corporate activities to foster sustainable finance by providing a clear definition of those corporate activities that can be named sustainable. In the United States, so far, such a regulation does not yet exist to this extent, but is debated (Christensen et al. 2021): e.g., in March 2022, SEC disseminated an initiative aiming to foster climate-related disclosures of domestic and foreign registrants in the registration statements and in the periodic reports (SEC 2022). Moreover, US companies seem to do a lot of voluntary disclosure (Cho et al. 2021). In China, from 2008 on, the disclosure of CSR activities is mandatory for certain companies. However, this does not require any behavioral changes from these companies, i.e., the regulation so far aims at transparency (Chen et al. 2018, Li et al. 2021). Further changes in regulation worldwide can be found in the Carrots & Sticks Report by Van der Lugt et al. (2020). In sum, the mentioned changes in regulation require profound information on the social and ecological activities. Standards like the GRI or international frameworks like the UN Global Compact already provide guidance to structure the publication of this information. Yet, these guidelines look at the different issues separately, like energy consumption, water use, and waste production. In contrast, TCA is a structured process, which can encompass the whole value chain of a company and thereby help to firmly anchor sustainability within the company and work out the connections between these individual issues. Thereby, it can both provide the necessary information for publication, and the necessary change of perspective to truly reach the aims of these regulating initiatives.

However, the relation between society/regulation, and companies is bidirectional. While companies are steered toward a sustainable transformation by this regulation, they need an environment that also allows them to undergo this transformation. Competition rules are of particular importance here. In order to be able to clearly articulate the adjustments required, companies must provide information that underpins their position. If a sustainable transformation is to succeed, externalities, e.g., with regard to the hitherto unpriced consumption of natural resources and their following impacts, must be integrated into pricing. However, a company can only do this if it is equally done by all competitors. Otherwise, competition will be distorted. TCA is able to present the costs of such externalities in a structured manner and thus provide the regulator with information for a fair regulation of competition.

Financiers. Related to the mentioned changes in regulations are changing perspectives of investors and banks. Particularly, developments in the European Union's regulation aim to ensure that financiers, especially investors and banks, more consciously incorporate social and environmental aspects into their investment and lending decisions. This is intended to channel financial flows toward a sustainable transformation of the economy. TCA is one building block to signal to financiers that a company is willing to bear its share in this transformation process and therefore is a reasonable investment or borrower.

In turn, companies wishing to undergo this sustainable transformation need financiers with a long-term perspective, as many of the changes, such as the development of sustainable products and production processes, the design of closed material cycles, and the transition to sustainable energy, take time. As a result, these processes are very rarely compatible with demands for high short-term returns. The signal effect of implementing TCA can make a company interesting, especially for those investors who aim for long-term investment in companies serious about striving for sustainable transformation.

Suppliers. Also, in future corporate decision-makers will require more information on social and ecological impacts to develop a *sustainable* supply chain. As Jasinski et al. (2015, p. 1125) state based on Bebbington (2007): "an assessment of an activity which is blind to upstream and downstream effects cannot fully address sustainability". Consequently, if companies strive to transform their business processes in a sustainable manner, then their efforts must not stop at their own factory gates. This view is also reflected in more recent regulation, resulting, for example, in the German Supply Chain Due Diligence Act (Gesetz über die unternehmerischen Sorgfaltspflichten in Lieferketten, LkSG; Bundesanzeiger Verlag 2021), which comes into force in 2023. Moreover, in February 2022 the European Commission issued a proposal for a Directive for Corporate Sustainability Due Diligence, which also aims at the monitoring of whole supply chains in terms of sustainability (European Parliament and Council 2022). As TCA is based on information that also incorporates resource consumption, pollution, and social impact of primary products finding their way into the production processes of the focal company, it can help to monitor activities of suppliers as wished for by regulation. In addition, it is also in the interest of many companies to have a better overview of the activities within their supply chain. For example, delivery failures and bottlenecks during the Corona pandemic showed how fragile these supply chains are. The implementation of TCA requires a more intensive examination of these interrelationships and can help to uncover weaknesses and hotspots by providing more transparency. However, this endeavor also is the most difficult in the context of TCA. So far transparency about the activities along the whole supply chain or more often supply network is difficult to achieve, particularly at the origin, and companies struggle to get truthful information. The implementation process of TCA can be the gateway to improve communication, receive the desired information, and trigger action toward more sustainability.

In return, suppliers expect reliable cooperation conditions and the focal company's willingness to support efforts toward more sustainable production and to accept resulting costs. Therefore, it is also in the interest of suppliers aiming at sustainable production processes and outcomes to enter a joint process of developing TCA with the focal company. The LCA, which forms an input to TCA as discussed in previous sections, provides valuable insights into the processes along the whole supply chain and helps to identify major drivers of a sustainable supply chain for all members of this supply chain.

Customers. Although customers, for example, in the luxury sector (Kunz et al. 2020), are not universally interested in the underlying production processes, an increasing critical attitude can be observed among certain customer groups. For example, more and more consumers are paying attention to so-called eco labels (Bastounis et al. 2021), which are supposed to guarantee compliance with certain social and/or ecological production standards. To obtain these labels, companies must disclose information about their sustainable activities. TCA can provide parts of this information and help to make companies' sustainability activities more visible.

Moreover, despite a long-standing debate about sustainable consumption, consumers are still unaware of their true resource consumption, as parts of the social and ecological costs are not included in the price tags of goods and services. Thus, awareness-raising can be an important objective for companies in the food industry or the fashion industry, for example. These industries have a particularly large environmental impact but are under strong price pressure at the same time. Affordable food prices, for example, play a major role in the overall social context, while the fashion industry, through the development of fast fashion, drives its customers to buy fast-changing, low-priced clothing. In order to achieve the transformation to more sustainable production envisaged by regulation, these industries also need the willingness of customers to understand and accept price increases. TCA can contribute to this by monetizing total resource consumption and its impacts and thus providing a basis for the aforementioned awareness-raising.

4.3 TCA as part of a sustainable management information and early detection system

Key issue: How can TCA be integrated with other sustainable information tools?

The implementation of TCA is an elaborated process that involves different departments of a company. In the present section, we show that this effort is worthwhile, as TCA can be combined with a number of other sustainable management instruments and thus a comprehensive sustainable management information and early detection system can be established. Over the last decades, a wide range of tools to provide information for a sustainable management has been developed. Thus, it is not possible to present all these instruments here. We therefore focus on those that are widely discussed and have a particularly close relationship to TCA. Moreover, we also concentrate on tools and do not refer to general developments in management, such as sustainable supply chain management, or comprehensive systems, such as environment or social management systems. Figure 4.2 illustrates the relation between TCA, the identified instruments, management processes, and interaction processes with the stakeholders mentioned in Section 4.2.

On the one hand, TCA can resort on input of three other instruments in the context of sustainable management: Life cycle analysis as discussed in Chapter 3, and additionally on substance flow analysis and carbon accounting. The earlier provides information about the flow of substances in a predefined system and thereby can inform about material that enters or leaves production processes (Huang et al. 2012). The latter is dedicated to the evaluation of the corporate carbon emissions (Schaltegger and Csutora 2012). It provides both physical data on these emissions and their monetary evaluation. Thereby, it covers an important part of ecological relevant resource consumption, which

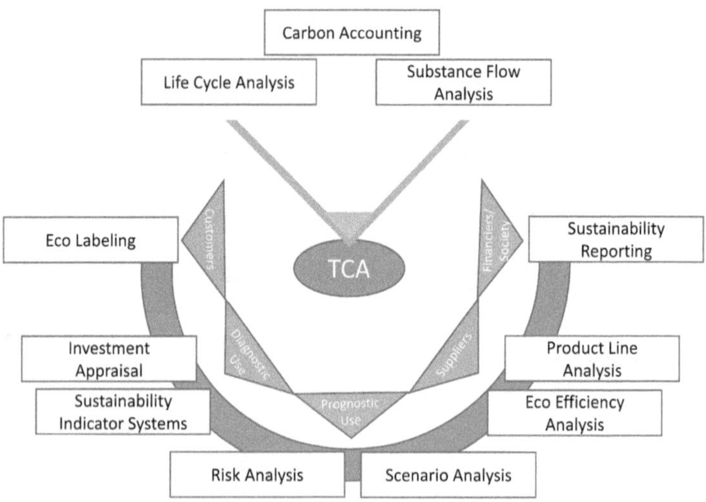

Figure 4.2 Relationships between TCA, identified instruments, management processes, and interaction processes with the stakeholders mentioned

has been neglected in classical cost accounting and provides data that is compatible with TCA.

On the other hand, TCA can provide input into a range of other sustainability management tools. It has to be mentioned that these tools also have relations with one another, but to keep the discussion comprehensible we just focus on their links to TCA. As indicated in Figure 4.2, the selected tools can be related to the previously mentioned stakeholders and to a company's management activities.

Related to the latter we differentiate between diagnostic and prognostic tools. Both sustainability indicator systems and investment appraisals are particularly helpful for internal decision processes in a diagnostic sense, i.e., providing information for operational decision processes. The Sustainability Balanced Scorecard (SBSC) constitutes one of the most prominent indicator systems. It is based on the Balanced Scorecard (BSC) developed by Kaplan and Norton (1992), which is part of an overarching strategy development and implementation process. The original form of this indicator system comprises four perspectives: The learning and employee-oriented, the process, the customer, and the financial perspective. Each perspective contains several indicators focusing on the inherent theme of the perspective. These indicators are intended to provide management with sufficient information to steer the company toward the predefined goals. If the values of the indicators deviate from preset targets, measures must be taken to achieve these target values and thereby implement the desired strategy. The SBSC enriches this original structure along several dimensions (Hansen and Schaltegger 2016). There are SBSCs in which sustainability indicators are integrated into each perspective, others have their own sustainability perspective, and others combine both. There are also cases where a classic BSC is used in parallel with a SBSC. The result of the TCA is a comprehensive monetary evaluation of the resource consumption and its impacts for the production of goods. This is related to the key figures of the financial perspective and could be integrated into it, e.g., to enable a comparison with purely economic costs and profit figures. In addition, further data are generated during the realization of the TCA, which can be integrated into the other perspectives. For example, the preceding LCA provides relevant information for the process perspective, as it can indicate (inefficient) changes in resource consumption over time, which, in turn, can initiate improvements to achieve more efficient production processes.

Decisions regarding investment alternatives are based on investment appraisals. In its classic version, these appraisals take only the economic perspective into account. An extension, which would include a sustainable perspective, requires an evaluation of the social and ecological costs. TCA provides a monetary evaluation of such costs. However, it should be noted that to date cash flow-based methods, such as the net present value method, are mainly used in investment appraisals (Vesty and Oliver 2014). Since TCA is based on a cost-oriented perspective, it does not provide cash flows but the monetary assessment of resource consumption and its impacts, which partly do not lead to any cash flows. Accordingly, the basic understanding behind the classic investment appraisal methods must be expanded, but the calculation methods are compatible since they all lead to monetary values.

TCA can also inform prognostic systems, in particular risk analysis and scenario analysis. Risk analysis is used to identify, evaluate, and weight risks relevant to the company. TCA can create added value, particularly with regard to the identification of risks since it looks at a wider range of resource uses than traditional corporate accounting systems. As a result, it sheds broader light on the resources needed for business activities, including those that do not currently generate economic costs but may do so in the future because, for example, regulation changes (Jasinski et al. 2015). Price tags on CO_2 consumption or fines for disregarding employee rights along the supply chain are examples of this.

Scenario analyses help to understand the future consequences of changing situational factors for the business model. In the sustainability context these analyses can cover both, the comparison of physical efficiency, like the carbon emissions in different scenarios (Säynäjoki et al. 2012), or the comparison of monetized values, like the true prices for food which is produced conventionally as compared to organically. The quality of scenario analysis depends on both the quality of input data and of the applied algorithms. TCA can add to the former reasonable monetized values.

By using Eco labels companies can create transparency for their customers about their sustainable efforts within their procurement and production processes (Tzilivakis et al. 2012). The different labels each require the provision of different social and environmental information. The implementation of a TCA leads to the generation of this information.

As previously mentioned, true sustainable management can only be achieved if companies also include suppliers into their considerations. Here, product line and eco efficiency analyses are of particular interest, as these are dedicated to the ecological and social impact generated along the whole value chain. The former comprises the assessment of products' life cycle impacts along ecological, social, and economic dimensions (Wang and Shen 2017). Thus, it is related to the previously mentioned LCA but incorporates a scenario perspective as it is implemented to compare different variants of a product to choose the most sustainable option. Consequently, both LCA as an antecedent of TCA, but also TCA itself can inform this comparison process.

The eco efficiency analysis provides a relation "between economic value creation and environmental impact added" (Möller and Schaltegger 2008, p. 78) and also takes a perspective across the entire value chain. Economic value can be operationalized by e.g., sales revenues. Environmental impact added can be indicated, for example, by land use, carbon emission, nitrogen input into soils, or water use as measured by LCA, or by monetized values of these impacts, which are provided by TCA.

Finally, as previously discussed, financiers and society demand increasing transparency about the degree of environmental and social sustainability in business practices. Accordingly, external reporting must be adapted and expanded, especially in the European Union due to its pioneering role. As already mentioned, a range of information is generated as part of the preparation for TCA that reflects the corporate impact on social and environmental areas. At the same time, this process provides important information for the required reporting. A joint implementation of TCA and redesigned reporting thus creates major synergy effects for the companies concerned, as external communication and internal management benefit simultaneously.

4.4 Practical ways to stepwise implement TCA

Key issue: Which expansion stages of a TCA implementation are reasonable?

The preceding paragraphs have shown that the introduction of TCA offers many advantages, as it can help overcome the challenges currently facing accounting, and as it is compatible with other tools of sustainable management. Nevertheless, it should be kept in mind that its implementation requires a certain amount of company resources and that this itself poses a number of challenges. This section therefore aims at showing what a step-by-step implementation could look like, what benefits even initial approaches without full implementation can have for companies, and what an institutional anchoring should look like. Figure 4.3 summarizes the following discussion.

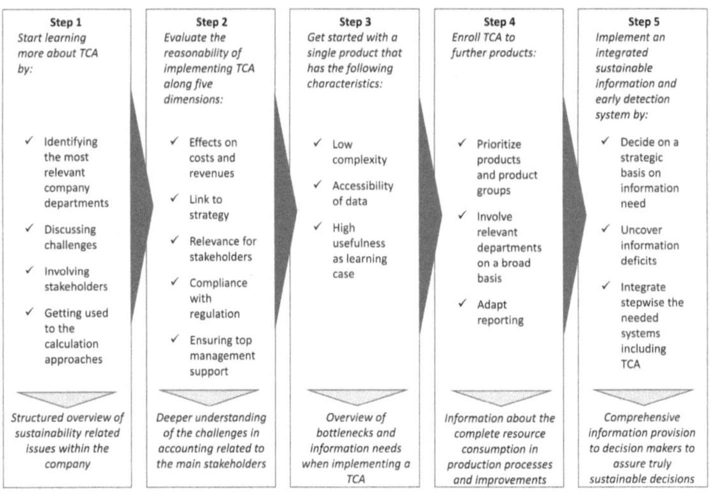

Figure 4.3 Summary of step-by-step implementation of TCA

Step 1: Start to learn more about TCA. For most accountants and companies TCA is a new cost calculation system, which only partly draws on well-known calculation approaches and existing data. Moreover, it carries with it several challenges that must be consciously addressed at the outset of any engagement with it. These challenges revolve around two major aspects, which are related to the calculation as such and its organizational embedding.

First, the definition and evaluation of externalities is a difficult task, as often there exist multiple reasonable ways of doing this, which results in rather a range of calculated values than one certain value (Bebbington and Larrinaga 2014). Therefore, Bebbington and Larrinaga (2014) argue that one single figure has less quality than a range of figures, which reflect the complexities and the uncertainty behind the TCA calculations. In this context, these authors also warn against uncritically aiming at mathematical precision for the calculation of such costs by citing Funtowicz and Ravetz (1994, p. 201). They, in turn, draw on Carl Friedrich Gauss by stating that a "[l]ack of mathematical culture is revealed nowhere so conspicuously as in meaningless precision in numerical calculation" (Bebbington and Larrinaga 2014, p. 407). In this sense, they argue for providing room for all stakeholders to discuss the choices made in the processes of TCA – or as they call it full cost accounting – as an indication of a high-quality costing process. Therefore, as mentioned, they argue for accepting a range of values rather than aiming at reaching one single value. This shows that TCA, despite its closeness to cost accounting, requires a fundamentally different perspective on the calculation of costs, as it deliberately allows for fuzziness in the calculation in order to better reflect reality. In classical accounting, such uncertainty is undesirable. Just think of the difficulties that can arise for the determination of tax payments, for example. The matter gets even more complicated since the transfer of physical phenomena to an accounting system can turn out to be difficult. Thus,

> [t]aken together, these issues make any full cost accounting exercise inherently contestable. Drawing on the previous discussion about sustainability science, we contend that the contestability of an account is not a limitation. Rather, it is a reality with which any account must work.
>
> (Bebbington and Larrinaga 2014, p. 406)

Consequently, in this first step, all parties involved must find clarity on what kind of costing system they are dealing with and how they should handle it in order to use it meaningfully.

Second, although TCA is a cost accounting system, it takes more than a cost accounting department to implement it. As mentioned, to perform TCA both physical quantities of resource consumption and its impacts are recorded and monetized. While monetary valuation is a core competence of accountants, the collection and processing of data on the physical level in production, sales, and along the entire supply chain is a competence of employees in these areas. TCA can therefore only be carried out on a cross-departmental basis and, since it also draws on data relating to upstream products, on a cross-company basis. The extent of the collaboration required depends on the scope of the TCA. For a trial costing of a single product or the introduction of a regular valuation of certain prestige products, the collaboration required is more manageable than for an implementation of TCA on a larger scale. Accordingly, even in this first stage of dealing with TCA, care should be taken to ensure that all parties who will be relevant in future are involved and agree on a common approach and understanding. Attention should be paid to possible resistance to TCA implementation at this stage and this should be taken seriously, otherwise, it will torpedo further steps as an example discussed in Herbohn (2015) illustrates. Among other things, the author elaborates on concerns about possible misuse of generated data, the exploitation of positions of power, and fears of management about strong resistance from certain stakeholders. Furthermore, the complexity of the calculation schemes is described as problematic, as several stakeholders do not understand them.

Consequently, companies should take their time to learn more about TCA before they start with the next step of implementation. This time is well spent, even if they conclude that a TCA is too complex or too expensive and should therefore not be implemented, because already at this stage an improved understanding of externalities and their influence on the degree of sustainability of corporate activities is developed, which can have an impact on everyday business.

Step 2: Evaluate the reasonability of implementing TCA. While the first step involves establishing a basic understanding of TCA and identifying the relevant stakeholders, the second step is to more precisely determine the added value of TCA for the company. Here, we refer back to the previous explanations on the challenges for accounting and the identified stakeholders. While the first step does not necessarily have to be characterized as an official project, a project should be set up at the latest in the herein discussed step and staffed with an interdisciplinary project team. Subsequently, the advantages and disadvantages of TCA for the company should be recorded in both quantitative and qualitative forms. A quantitative evaluation comprises, for example, expected set-up and maintenance costs related to the implementation and operation of TCA, as well as cost savings due to the identification of, for example, inefficient processes, unnecessary waste, or illness days related to poor social standards. Also, possible positive and negative effects on revenues can be estimated in case that the implementation of TCA is communicated to customers for marketing purposes or that true costs are applied when setting product or service prices. Qualitative data, in contrast, are needed to learn more about the relevance of the aforementioned challenges for accounting in the focal company and the possibilities offered by a TCA system tailored to that company for dealing with them. Specifically, this means examining how the company is affected by changes in the regulation of reporting and supply chains, how it depends on investors who demand sustainability-oriented information, and whether this information is also important for its relationship with customers. Above all, however, the project team must determine the extent to which TCA is compatible with the actual corporate strategy, i.e., whether there is actually an interest in sustainable transformation and the resulting transformation of management. If this is not the case, one of the most important components for implementing such a project will be missing: Top management support.

While the first phase comprises a rather unstructured engagement with TCA, this phase involves a more detailed handling and definition of it and a structured analysis of the advantages and disadvantages of TCA. This also leads to more consciousness about the mentioned challenges of accounting and possible re-calibrations in dealing with relevant stakeholders. This increase in knowledge creates transparency about these challenges affecting the company, which must be addressed even without a further implementation of TCA.

Step 3: Get started with a single product. If the company concludes that it makes sense to implement TCA, the data generation and calculation processes can be set up. However, as TCA is a complex calculation procedure, it is advisable to start with a manageable costing object, i.e., the project team should identify a single product that is suitable for a pilot costing project. Suitability is determined by complexity and data availability. For a first attempt to implement TCA, a product of low complexity is more reasonable than a highly complex one, as the latter requires much more data and calculation steps and would possibly lead to even higher uncertainties hardly communicable to stakeholders. Moreover, good access to required data accelerates TCA processes and leads to faster results. To assess the required information, system boundaries for the product have to be defined and relevant information has to be identified. Thereafter, all departments, which have relevant data but are not yet part of the project, must be convinced to provide the needed information. When the necessary information is compiled, the project team can assess the true costs according to the method described in Chapter 3. This can form the basis for discussion processes about possibilities for environmental or social improvements considering this pilot product. But even more essential than the calculation of this final value is the process of implementing an exemplary TCA itself. The pilot product can be used as a learning case to predict the challenges arising with more complex products and to estimate the resources required for a further expansion of TCA.

Thus, the most important goal of this phase is not to obtain a correct value for the true costs of the selected product, but to build up experience regarding the costing process and the existing bottlenecks and information needs. This enables designing the extension of TCA to entire product groups more efficiently. In sum, this step is still part of a learning phase and does not yet provide a fully developed management information system.

Step 4: Enroll TCA to further products. The experience gained from the third step can be used to carry out an extension of TCA to product groups and broader areas and hence finally to obtain a management information system. To achieve this goal efficiently, products or product groups should be prioritized, allowing for a stepwise enrollment. In this stage, the project team should initiate processes to adapt the existing corporate information systems to the needs of TCA and to implement a regular collection of necessary information, as the relevant data should be collected in a structured manner and integrated in regular reporting activities. Thus, the affected departments, like accounting, sustainability management, production, logistics, marketing, or supply chain management, have to be involved on a broader scale going beyond single data inquiries. In essence, this phase comprises the establishment of standard routines and procedures to embed TCA in the processes related to management information provision.

The result of this step is a management information system that provides decision-makers with information about the true resource consumption and its impacts and allows insights into bottlenecks and potential for improvements by uncovering hidden costs, which, for example, result from resource waste or working conditions that miss social standards. However, beyond this economic perspective on the processes, the gained information also helps decision-makers to better understand the current status of sustainability in their company and, thus, opens up perspectives for improvements along environmental and social dimensions.

Step 5: Implement an integrated sustainable information and early detection system. In the final step, TCA can be enlarged to an integrated sustainable information and early detection system. To achieve this goal, management must first determine at the strategic level which sustainability-oriented information is relevant in connection with which stakeholders. The next step is to compare the current provision of information with the actual need for information. To do this, systems, including TCA, must be identified in the company that already provide information for sustainable management. These can be found in all areas of the company and are therefore not specific to accounting. If a lack of information is identified, supplementary instruments must be considered, such as an SBSC to provide structured sustainability-oriented key figures. Subsequently, the existing systems have to be connected and further systems have to be integrated.

This step requires major effort by large parts of the organization and can therefore only be implemented by companies with strong resources. However, this step develops the basic idea behind the discussion of management control systems in the direction of sustainability. It can therefore also be implemented in a simplified form, e.g., by integrating only one existing system with the newly introduced TCA. Such a simplified approach already enables a more comprehensive view of a company along the ecological and social dimension in addition to the economic one.

4.5 Review questions

1 Why is TCA a natural progression of classical accounting?
2 Discuss the challenges that accounting is currently facing.
3 Name three tools to provide information on sustainability activities in the corporate context and discuss their relation to TCA.
4 What are the major stakeholders to be considered when implementing TCA?
5 Why can engaging with TCA also lead to benefits, even if a company decides against adoption afterwards?
6 Discuss whether TCA is an MA or FA tool.
7 What advantages can a company gain from implementing TCA, even if the calculated true costs are not fully included in the products' or services' pricing?
8 What are the most important challenges in implementing the five implementation steps mentioned above?

References

Bastounis, A., Buckell, J., Hartmann-Boyce, J., Cook, B., King, S., Potter, C., Bianchi, F., Rayner, M., & Jebb, S. A. (2021). The impact of environmental sustainability labels on willingness-to-pay for foods: A systematic review and meta-analysis of discrete choice experiments. *Nutrients, 13*(8), 2677.

Bebbington, J. (2007). *Accounting for sustainable development performance*. CIMA: Oxford.

Bebbington, J. (2021). The Foundations of Environmental Accounting. In: Bebbington, J., Larrinaga, C., O'Dwyer, B., & Thomson, I. (eds) *Routledge Handbook of Environmental Accounting*. Routledge, Electronic version: Chapter 2.

Bebbington, J., & Larrinaga, C. (2014). Accounting and sustainable development: An exploration. *Accounting, Organizations and Society, 39*(6), 395–413.

Bebbington, J., & Thomson, I. (2013). Sustainable development, management and accounting: Boundary crossing. *Management Accounting Research, 24*, 277–283.

Bundesanzeiger Verlag. (2021). *Gesetz über die unternehmerischen Sorgfaltspflichten zur Vermeidung von Menschenrechtsverletzungen in Lieferketten (Lieferkettensorgfaltspflichtengesetz – LkSG)*. Bundesanzeiger. Available at: https://wirtschaft-entwicklung.de/fileadmin/user_upload/5_Wirtschaft_und_Menschenrechte/Downloads/Lieferkettensorgfaltspflichtengesetz.pdf (Accessed October 27, 2022)

Burritt, R. L. (2005). Challenges for Environmental Management Accounting. In: Rikhardsson, P. M., Bennett, M., Bouma, J. J., & Schaltegger, S. (eds) *Implementing Environmental Management Accounting: Status and Challenges* (19–44). Springer.

Chen, Y.-C., Hung, M., & Wang, Y. (2018). The effect of mandatory CSR disclosure on firm profitability and social externalities: Evidence from China. *Journal of Accounting and Economics, 65*, 169–190.

Cho, C. H., Krasodomska, J., Ratliff-Miller, P., & Godawska, J. (2021). Internationalization and CSR reporting: Evidence from US companies and their Polish subsidiaries. *Meditari Accountancy Research, 29*(7), 135–162.

Christensen, H. B., Hail, L., & Leuz, C. (2021). Mandatory CSR and sustainability reporting: Economic analysis and literature review. *Review of Accounting Studies*, *26*, 1176–1248.

Endenich, C. (2014). Economic crisis as a driver of management accounting change. *Journal of Applied Accounting Research*, *15*(1), 123–149.

European Commission. (2001). *Green paper: Promoting a European Framework for Corporate Social Responsibility*. Available at: https://ec.europa.eu/commission/presscorner/detail/en/DOC_01_9 (Accessed October 27, 2022)

European Commission. (2006). *Implementing the partnership for growth and jobs: making Europe a pole of excellence on corporate social responsibility*. Available at: https://eur-lex.europa.eu/LexUriServ/LexUriServ.do?uri=COM:2006:0136:FIN:EN:PDF (Accessed October 27, 2022)

European Commission. (2011). *A renewed EU strategy 2011-14 for corporate social responsibility*. Available at: https://www.europarl.europa.eu/meetdocs/2009_2014/documents/com/com_com(2011)0681_/com_com(2011)0681_en.pdf (Accessed October 27, 2022)

European Commission. (2022). *Provisional agreement for a directive of the European Parliament and of the Council amending Directive 2013/34/EU, Directive 2004/109/EC, Directive 2006/43/EC and Regulation (EU) No 537/2014, as regards Corporate Sustainability Reporting*. Available at: https://www.consilium.europa.eu/media/57644/st10835-xx22.pdf (Accessed October 9, 2022)

European Parliament and Council. (2013). *Directive 2013/34/EU of the European Parliament and of the Council of 26 June 2013 on the annual financial statements, consolidated financial statements and related reports of certain types of undertakings, amending Directive 2006/43/EC of the European Parliament and of the Council and repealing Council Directives 78/660/EEC and 83/349/EEC*. Available at: https://eur-lex.europa.eu/legal-content/EN/TXT/PDF/?uri=CELEX:32013L0034&from=DE (Accessed October 27, 2022)

European Parliament and Council. (2014). *Directive 2014/95/EU of the European Parliament and of the Council of 22 October 2014 amending Directive 2013/34/EU as regards disclosure of non-financial and diversity information by certain large undertakings and groups*. Available at: https://eur-lex.europa.eu/legal-content/EN/TXT/PDF/?uri=CELEX:32014L0095 (Accessed August 23, 2022)

European Parliament and Council. (2019). *Regulation (EU) 2019/2088 of the European Parliament and of the Council of 27 November 2019 on sustainability-related disclosures in the financial services sector*. Available at: https://op.europa.eu/en/publication-detail/-/publication/4f50e277-1a53-11ea-8c1f-01aa75ed71a1 (Accessed August 23, 2022)

European Parliament and Council. (2020). *Regulation (EU) 2020/852 of the European Parliament and of the Council of 18 June 2020 on the establishment of a framework to facilitate sustainable investment, and amending Regulation (EU) 2019/2088*. Available at: https://op.europa.eu/en/publication-detail/-/publication/e5ba36a8-b454-11ea-bb7a-01aa75ed71a1/language-en/format-PDFA2A (Accessed August 23, 2022)

European Parliament and Council. (2022). *Proposal for a Directive of the European Parliament and of the Council on Corporate Sustainability Due Diligence and amending Directive (EU) 2019/1937*. Available at: https://ec.europa.eu/info/sites/default/files/1_1_183885_prop_dir_susta_en.pdf (Accessed October 6, 2022)

Funtowicz, S. O., & Ravetz, J. R. (1994). The worth of a songbird: Ecological economics as a post-normal science. *Ecological Economics*, *10*(3), 197–207.

Giroux, G. (2017). *Accounting History and the Rise of Civilization* (Vol. I). Business Expert Press.

Gray, R. (2002). The social accounting project and *accounting, organizations and society* – Privileging engagement, imaginings, new accountings and pragmatism over critique? *Accounting, Organizations and Society*, *27*, 687–708.

Gray, R. (2010). A re-evaluation of social, environmental and sustainability accounting: An exploration of an emerging trans-disciplinary field? *Sustainability Accounting, Management and Policy Journal*, *1*(1), 11–32.

Hansen, E. G., & Schaltegger, S. (2016). The sustainability balanced scorecard: A systematic review of architectures. *Journal of Business Ethics*, *133*, 193–221.

Herbohn, K. (2015). A full cost environmental accounting experiment. *Accounting, Organizations and Society*, *30*, 519–536.

Hines, R. (1991). On valuing nature. *Accounting, Auditing and Accountability Journal*, *4*(3), 27–29.

Huang, C. L., Vause, J., Ma, H. W., & Yu, C. P. (2012). Using material/substance flow analysis to support sustainable development assessment: A literature review and outlook. *Resources, Conservation and Recycling*, *68*, 104–116.

IMA. (2019). *IMA Management Accounting Competency Framework*. Institute of Management Accountants.

Jasinski, D., Meredith, J., & Kirwan, K. (2015). A comprehensive review of full cost accounting methods and their applicability to the automotive industry. *Journal of Cleaner Production*, *108*, 1123–1139.

Kaplan, R. S., & Norton, D. P. (1992). The balanced scorecard. Measures that drive performance. *Harvard Business Review*, *70*, 71–79.

Kunz, J., May, S., & Schmidt, H. J. (2020). Sustainable luxury: Current status and perspectives for future research. *Business Research*, *13*(2), 541–601.

Li, Z., Zou, F., & Mo, B. (2021). Does mandatory CSR disclosure affect enterprise total factor productivity? *Economic Research-Ekonomska Istraživanja*, *35*(1), 4902–4921.

Mathews, M. R. (1997). Twenty-five years of social and environmental accounting research. *Accounting, Auditing & Accountability Journal*, *10*(4), 481–531.

McKernan, J. F. (2007). Objectivity in accounting. *Accounting, Organizations and Society*, *32*(1–2), 155–180.

Milne, M. J. (1996). On sustainability; The environment and management accounting. *Management Accounting Research*, *7*(1), 135–161.

Möller, A., & Schaltegger, S. (2008). The sustainability balanced scorecard as a framework for eco-efficiency analysis. *Journal of Industrial Ecology*, *9*(4), 73–83.

Owen, D. (2008). Chronicles of wasted time? A personal reflection on the current state of, and future prospects for, social and environmental accounting research. *Accounting, Auditing & Accountability Journal*, *21*(2), 240–267.

Säynäjoki, A., Hinonen, J., & Junnila, S. (2012). A scenario analysis of the life cycle greenhouse gas emissions of a new residential area. *Environmental Research Letters*, *7*, 1–10.

Schaltegger, S., & Burritt, R. L. (2010). Sustainability accounting for companies: Catchphrase or decision support for business leaders? *Journal of World Business*, *45*, 375–384.

Schaltegger, S., & Csutora, M. (2012). Carbon accounting for sustainability and management. Status quo and challenges. *Journal of Cleaner Production*, *36*, 1–16.

Securities and Exchange Commission (SEC). (2022). *The enhancement and standardization of climate-related disclosures for investors*. Available at: https://www.sec.gov/rules/proposed/2022/33-11042.pdf (Accessed March 28, 2022)

Tschopp, D., Wells, S., & Barney, D. (2012). The institutional promotion of corporate social responsibility reporting. *Journal of Academic and Business Ethics*, *6*, 1–18.

Tzilivakis, J., Green, A., Warner, D., McGeevor, K., & Lewis, K. (2012). A framework for practical and effective eco-labelling of food products. *Sustainability Accounting, Management and Policy Journal*, *3*(1), 50–73.

Van der Lugt, C. T., van de Wijs, P. P., & Petrovics, D. (2020). *Carrots & Sticks 2020 - Sustainability reporting policy: Global trends in disclosure as the ESG agenda goes mainstream*. Global Reporting Initiative (GRI) and the University of Stellenbosch Business School (USB).

Vesty, G., & Oliver, J. (2014). Corporate strategy and accounting for sustainability in investment appraisal. *Corporate Ownership & Control*, *11*(2), 377–388.

Wang, L., & Shen, B. (2017). A product line analysis for eco-designed fashion products: Evidence from an outdoor sportswear brand. *Sustainability*, *9*, 1–12.

Zvezdov, D., & Schaltegger, S. (2013). Sustainability Accounting. In: Idowu, S.O., Capaldi, N., Zu, L., & Gupta, A.D. (eds) *Encyclopedia of Corporate Social Responsibility* (2363–2370). Springer.

5 True Cost Accounting in Practice: Case Studies

Abstract
This chapter examines the evolution and impact of True Cost Accounting (TCA) through various case studies, showcasing the progression from basic carbon offsetting to comprehensive sustainability initiatives. The journey begins with companies focusing on carbon neutrality, such as Atmosfair and ClimatePartner, highlighting their approaches to CO_2 compensation. We then explore organizations extending their efforts to broader environmental and social dimensions, including Ecolabels like Bioland and social initiatives exemplified by the WeCare label. Furthermore, we delve into the participatory consumer model of "Du bist hier der Chef!". The chapter culminates with innovative approaches that integrate TCA directly into retail and consumer engagement, particularly through the approach of EOSTA and the nationwide PENNY true cost campaign in German food retail 2023. These examples illustrate the increasing depth and scope of TCA applications, ultimately emphasizing its potential to drive sustainable business practices and informed consumer choices.

Key takeaways

After reading this chapter, you will:

- Understand the different levels of TCA application, from carbon offsetting to comprehensive sustainability initiatives
- Gain insights into how companies like Atmosfair and ClimatePartner implement CO_2 compensation strategies
- Recognize the significance of environmental and social labels such as Bioland and WeCare in promoting sustainable practices
- Learn about the participatory approach of "Du bist hier der Chef" in enhancing transparency and fairness in the food supply chain
- Discover innovative TCA applications in retail, with a focus on EOSTA and the PENNY case of 2023, which integrate true cost principles directly with consumer pricing
- See the potential for TCA to drive global sustainability practices and foster informed consumer behavior

DOI: 10.4324/9781003345190-8

5.1 Atmosfair: Fostering Climate Conscious Travel through TCA

Key issue: What is Atmosfair, how does it work, and how is it connected to True Cost Accounting?

"Atmosfair" leverages True Cost Accounting (TCA) to provide a comprehensive assessment of the environmental costs associated with air travel, focusing on carbon offsetting through renewable energy projects (Atmosfair 2022, p. 13).

Atmosfair is a climate protection organization specializing in travel-related carbon offsets. With a primary focus on the aviation sector, Atmosfair addresses one of the most challenging areas for carbon reduction, given the current lack of feasible low-emission alternatives (Atmosfair 2022, p. 8 ff.). By facilitating the compensation of greenhouse gasses through investments in renewable energy projects, Atmosfair plays a pivotal role in the global effort to mitigate climate change (BMU 2017). The organization's relevance extends internationally as it offers travelers worldwide the means to offset their carbon footprint, thus promoting more sustainable travel habits (Atmosfair 2021). Additionally, Atmosfair's projects are often certified by the "Gold Standard CER", ensuring that they not only improve climate impacts but also provide social and biodiversity benefits (Umweltbundesamt 2020, Atmosfair 2022).

Atmosfair operates under the principle that carbon offsetting can be a viable climate action, provided that certain conditions are met. These include the absence of less carbon-intensive alternatives, the potential for technological advancements toward low-emission products, and the comprehensive accounting of emissions from the product or activity in question (Atmosfair 2022).

The methodology involves:

- **Emission calculation**: Calculating the carbon emissions of flights, considering factors such as aircraft type, flight distance, and load factors (Atmosfair 2023).
- **Offset projects**: Compensating these emissions by investing in high-quality, certified renewable energy projects in developing countries. These projects, like wind farms and efficient cookstoves, help reduce CO_2 emissions and support sustainable development (Atmosfair 2024a, 2024b).

Atmosfair ensures high transparency and credibility by meticulously tracking and reporting the performance and impact of their projects (Atmosfair 2021, p. 39). However, it is important to note that these carbon calculators often exclude some emissions, which can also be significant (Atmosfair 2023).

Critique and Additional Considerations

- Is it truly climate-neutral?: While offsets are beneficial, they do not make flights completely climate-neutral. True climate neutrality would involve avoiding the flight altogether (Sacchi et al. 2023). Additionally, there are risks associated with some offset projects, such as reforestation efforts where new forests could potentially burn and release stored CO_2 back into the atmosphere (Lefebvre et al. 2021).

- Choosing the right agency: For effective offsetting, consumers should look for projects certified by the "Gold Standard CER". Atmosfair, along with Klimakollekte and Primaklima, received top ratings from "Stiftung Warentest" (Germany's leading consumer testing organization) for their high-quality projects that include additional social and biodiversity benefits (Spiegel Online 2020).

Atmosfair's model aligns with the TCA framework by internalizing the external costs of carbon emissions. By converting the environmental cost into monetary terms and investing these funds into renewable energy (Atmosfair 2022), Atmosfair provides a clear example of how TCA can guide sustainable practices in the travel industry. The organization's focus on high transparency and the rigorous documentation of their projects ensure that the TCA principles are upheld.

The evaluation of Atmosfair's approach demonstrates that while carbon offsetting is not a panacea, it is a critical interim measure in the broader strategy for achieving carbon neutrality in sectors where immediate alternatives are not yet viable. Although the Atmosfair approach is commended and there are calls for it to be used in other areas, it is also criticized for being "business with a guilty conscience" (cf. Gojdka and Timmler 2019). In air travel in particular, a long-term solution would probably be a direct tax on kerosene emissions, a ban on domestic flights in Germany or the incentivization of rail travel and the expansion of routes with night trains in Europe (see Umweltbundesamt 2021). The approach of Atmosfair should therefore only represent a transition to the actual transformation of mobility in the context of flying.

5.2 ClimatePartner: Enabling corporate climate neutrality through TCA

Key issue: What is ClimatePartner, how does it work, and how is it connected to True Cost Accounting?

"ClimatePartner" integrates TCA principles by offering companies a structured pathway to achieving climate neutrality, emphasizing comprehensive emissions calculations and carbon offsetting.

ClimatePartner provides businesses with the tools and strategies to achieve climate neutrality. By calculating a company's total greenhouse gas emissions and offering offset solutions through various climate protection projects, ClimatePartner addresses the growing demand for corporate environmental responsibility. This service is vital for companies worldwide looking to reduce their carbon footprint and enhance their sustainability credentials (ClimatePartner 2024a). ClimatePartner's impact is globally significant, as it helps businesses from diverse sectors and regions contribute to the fight against climate change. However, it is crucial to note that the term "climate-neutral" on products was not legally protected, and therefore, may not always reflect the full extent of emissions being offset or the environmental efforts undertaken by the manufacturer (Braatz 2022). However, the European Union has meanwhile described the following in a directive:

> Such statements should be prohibited under all circumstances and included in the list in Annex I of Directive 2005/29/EC, as consumers are misled into believing that these statements pertain to the product itself or to its provision and production, or that the consumption of this product has no impact on the environment. Examples of such statements include "climate neutral", "certified CO_2-neutral", "CO_2-positive", "with climate compensation", "climate friendly", and "with a reduced CO_2 footprint".

ClimatePartner's approach involves a detailed assessment of a company's entire supply chain to calculate its total carbon emissions (ClimatePartner 2024a, p. 18). This comprehensive analysis ensures that all indirect and direct emissions are accounted for. The company then provides offset solutions, which include projects like biogas, biomass, clean cookstoves, solar energy, and forest conservation (ClimatePartner 2024b). These projects are meticulously selected and monitored to ensure they meet high standards of effectiveness and transparency. ClimatePartner's methodology aligns with TCA by ensuring that the environmental costs are fully accounted for and offset in a way that promotes sustainable development (ClimatePartner 2024b, 2024c).

However, recent critiques highlight significant issues with carbon offsetting. For instance, a report by Foodwatch points out that "climate-neutral" labels on products do not necessarily guarantee their environmental friendliness, as they might not cover all emissions or reflect genuine efforts to reduce them (Foodwatch 2022). Additionally, concerns about the effectiveness of certain offset projects, especially forest conservation, have been raised. Studies have shown that many of these projects fail to deliver the claimed emission reductions, with only a small percentage actually preventing deforestation and CO_2 emissions (West et al. 2023).

ClimatePartner's services exemplify the application of TCA in a corporate context by transforming traditional accounting practices to include environmental costs. By helping companies understand and mitigate their environmental impact, ClimatePartner advances the implementation of TCA principles. The evaluation of ClimatePartner's approach highlights its strengths in providing practical, scalable solutions for companies aiming for climate neutrality.

However, the potential weaknesses include ensuring the long-term success and verification of the offset projects. For example, forest conservation projects often rely on speculative predictions about future deforestation, which can lead to overestimations of the emissions they actually prevent. Furthermore, the notion of "climate neutrality" can sometimes mask the real issue, encouraging businesses and consumers to believe that a small financial contribution can completely offset their carbon footprint without making substantial changes to their practices (Foodwatch 2022).

The danger of greenwashing is significant, as noted by experts from the New Climate Institute, who argue that labeling changes alone do not address the fundamental flaws in how offsets are calculated and marketed (NewClimate Institute 2024). Moreover, a study by Probst et al. (2023) has revealed that only about twelve percent of carbon offset certificates effectively contribute to CO_2 reduction, highlighting the inefficiency and potential deception in current offset practices.

Overall, while ClimatePartner's model demonstrates the feasibility and necessity of integrating TCA into corporate strategies, it is essential to approach carbon offsetting with a critical eye. Effective climate action should focus not only on compensating emissions but also on substantial reductions and systemic changes to avoid greenwashing and ensure genuine environmental benefits.

5.3 EU Organic Label

Key issue: What is the EU Organic Label, how does it work, and how is it connected to True Cost Accounting?

The EU Organic Label is a certification system promoting sustainable agricultural practices. However, it does not implement a comprehensive TCA methodology. It aims to ensure ecological and social standards in farming (EU 2018a).

The EU Organic Label is one of the most recognized and widespread organic labels globally. It marks products produced in accordance with European regulations for organic farming, adhering to stringent guidelines that limit the use of pesticides, fertilizers, and genetically modified organisms (GMOs) (EU2018b). The label holds international significance as a standard for organic products, recognized both within the EU and worldwide.

While the EU Organic Label does not directly follow a TCA approach, it indirectly supports TCA principles through stringent ecological standards. The methodology includes:

- **Strict environmental regulations:** Limiting chemical pesticides and synthetic fertilizers, promoting soil fertility, and ensuring animal welfare (EU 2018a).
- **Annual inspections:** Farms are inspected at least once a year to ensure compliance with standards (EU 2018c).
- **Labeling requirements:** Products with less than 95% organic ingredients cannot bear the label (EUCOM, 2024a).

Despite these measures, there are TCA-related shortcomings:

- **No comprehensive TCA:** It does not account for all external costs associated with agricultural production (NDR 2022).
- **Limited transparency:** Origin information is not always fully transparent, especially in mixed products (Lebensmittelklarheit 2022, EUCOM 2024b).

The EU Organic Label contributes significantly to promoting sustainable agricultural practices and offers consumers a guide to purchasing environmentally friendly products. However, it does not meet all the criteria of comprehensive TCA. The benefits of the label lie in promoting organic farming and improving food quality, which, from the point of view of the TCA methodology, have a positive effect on externalities compared to conventional agriculture (Michalke et al. 2023). TCA approaches show that organic farming practices usually cause lower externalities. A study by Oebel et al. (2024) analyzed this and showed that an adjustment of the VAT would not only reduce externalities in the agricultural sector and expand organic farming but would also have a positive impact on the national budget. Nonetheless, integrating all external costs remains a challenge that needs to be addressed to reveal the true ecological and social costs of food production.

5.4 Bioland

Key issue: What is Bioland, how does it work, and how is it connected to True Cost Accounting?

"Bioland" is a German organic farming association known for its high standards. The association integrates aspects of TCA by incorporating environmental and social criteria into its guidelines. Consequently, the same classification regarding TCA applies for Bioland as for the EU Organic Label (cf. Section 6.3).

Bioland is one of the leading organic farming associations in Germany, with over 7000 members. The Bioland guidelines exceed the EU organic regulations, covering comprehensive environmental and social standards. Bioland products are recognized not only in Germany but also internationally, representing high organic quality (Bioland 2024a). In the meantime, Bioland has also concluded contracts with large German food retailers such as Lidl, so that the products are now also available in large quantities in the German discount segment.

Bioland actively supports farming practices aligned with TCA principles. The methodology includes:

- **No genetic engineering:** Strict prohibition of GMOs.
- **Site selection:** Considering environmental contamination at site selection and regularly testing soil and food for pollutants.
- **Biodiversity:** Annual biodiversity enhancement contributions by members to promote species diversity.

- **Renewable energy:** Efficient use and promotion of renewable energy sources.
- **Social responsibility:** Commitment to social standards and human rights.

Bioland significantly contributes to the spread of sustainable farming practices, setting high standards for its members. These stringent standards are reflected in the quality of Bioland products, which often outperform conventional and EU organic products when external costs are considered. (European Network for Rural Development 2022, Bioland 2024b)

Bioland demonstrates how integrating environmental and social criteria into agricultural production can contribute to comprehensive sustainability and serves as a model for other farming associations and certification systems worldwide (European Network for Rural Development 2022, Bioland 2024b). Nevertheless, like the EU Organic Label, this approach is not a comprehensive approach in line with all TCA indicators.

5.5 We Care: The seal for sustainable corporate and supply chain management

Key issue: What it "We Care", how does it work, and how is it connected to True Cost Accounting?

"We Care" is a certification system that integrates TCA principles within the food industry, emphasizing comprehensive management of ecological and social impacts throughout the entire value chain. This initiative promotes sustainable business practices by ensuring that companies are responsible for their environmental and social footprints (We Care 2024).

We Care targets food industry businesses worldwide, offering a robust certification standard that addresses entire product ranges and incorporates sustainability at all operational levels and within supply chains. *We Care* is managed by the Research Institute of Organic Agriculture (FiBL) Germany e.V., which holds all rights to the standard and the associated We Care seal.

Given the global interconnectedness of food production and distribution, *We Care's* standards are relevant internationally, fostering sustainable practices across various markets and cultures. This global applicability helps companies align with international sustainability goals and respond to the growing demand for transparent and responsible business operations from consumers and stakeholders.

We Care employs TCA by requiring companies to comprehensively evaluate and monetize their environmental and social impacts. The certification process involves several critical steps:

- **Initial assessment:** Companies undergo a thorough evaluation to identify current sustainability practices and areas for improvement, collecting primary and secondary data on various sustainability indicators.

- **Implementation of sustainability practices:** Based on the assessment, companies implement necessary changes to enhance their sustainability performance. This might include reducing greenhouse gas emissions, improving labor conditions, or ensuring fair trade practices throughout the supply chain.
- **External verification:** Independent auditors review the company's sustainability practices and data. This verification ensures that the reported practices and impacts are accurate and credible.
- **Certification:** Upon successful verification, companies receive the *We Care* certification, which can be displayed on products and used in corporate communications (We Care 2023).

The *We Care* methodology is designed to be both comprehensive and flexible, allowing integration of existing certifications (such as organic or fair trade) while addressing gaps in sustainability management. Certification levels – basic and advanced – cater to companies at different stages of their sustainability journey, encouraging continuous improvement.

The *We Care* standard encompasses four main action areas:

- **Corporate governance:** Ensuring companies adhere to sustainable management practices and principles.
- **Supply chain management:** Promoting sustainability within supply chains, encouraging the use of producer-oriented certifications.
- **Environmental management:** Implementing practices that reduce environmental impact and promote resource efficiency.
- **Employee responsibility:** Ensuring fair and ethical treatment of workers, including safe working conditions and fair wages (We Care 2023).

We Care aligns with TCA principles by promoting a holistic view of sustainability, encompassing environmental, social, and economic impacts. This approach ensures that companies not only measure their financial performance but also account for their broader impacts on society and the environment (We Care 2023).

The integration of independent verification and comprehensive sustainability management makes *We Care* a robust and credible certification system. It supports transparency and accountability, enabling companies to effectively communicate their sustainability efforts to stakeholders.

In conclusion, *We Care* exemplifies the application of TCA in the food industry, providing a comprehensive framework for managing and verifying sustainability impacts. By fostering continuous improvement and integrating existing standards, *We Care* helps companies achieve and communicate their sustainability goals, enhancing their credibility and competitiveness in the global market. This aligns with our understanding of TCA, which seeks to provide a complete picture of a company's value creation and impact, ultimately leading to more sustainable and equitable business practices.

5.6 "Du bist hier der Chef": Empowering consumers through True Cost Accounting

Key issue: What is "Du bist hier der Chef", how does it work, and how is it connected to True Cost Accounting?

"Du bist hier der Chef (DBHDC)" integrates TCA principles by involving consumers directly in the decision-making process for the pricing and quality of food products (Grohmann et al. 2023), ensuring transparency and fair compensation for farmers (Abel 2020).

DBHDC is a consumer initiative inspired by the French movement "C'est qui le patron?!" launched in 2016 (Abel and Bätzing 2020). The initiative aims to create transparency in the quality and price of food products while guaranteeing fair remuneration for farmers. Founded in June 2019 by Nicolas Barthelmé and eight other consumers (Du bist hier der Chef 2024), DBHDC seeks to empower consumers to make informed choices about the products they purchase, contributing to a more sustainable and equitable food system (Abel 2020). The initiative's international relevance is underscored by its success in replicating the French model in Germany, helping farmers receive a fair price for their products (Du bist hier der Chef 2024) and enabling consumers to directly influence the agricultural market.

DBHDC operates under the TCA principle by calculating the true costs associated with food production, including environmental and social factors. The methodology involves creating consumer questionnaires to determine the preferences and willingness to pay for various product attributes. For instance, the first questionnaire launched in December 2019 focused on milk. It resulted in a high participation rate and culminated in the introduction of a so-called "consumer-driven milk" in July 2020 (Leopold 2020). This milk, produced by 15 farms and available in five regional listings, set one of the highest milk prices in Germany, reflecting the true cost of sustainable dairy farming (Du bist hier der Chef 2021).

The initiative continues to expand its product range through similar consumer surveys, with the egg survey completed in October 2020 and the potato survey completed in April 2022 (Du bist hier der Chef 2024). DBHDC's rigorous process ensures that consumer preferences directly shape product specifications and pricing, promoting fair compensation for farmers and sustainability in food production (Grohmann et al. 2023)

DBHDC exemplifies the integration of TCA in the food industry by incorporating the hidden costs of food production into the pricing and decision-making process. By allowing consumers to decide on product attributes and pricing, DBHDC ensures that the environmental and social costs are accounted for, promoting transparency and fairness. While the concept of "climate neutrality" has faced some criticism for potentially misleading claims (Fischer and Knuth 2023), DBHDC's focus on consumer participation and transparency offers a promising model for achieving sustainable and equitable food production.

5.7 EOSTA: Transparency for fruits and vegetables

Key issue: What is EOSTA, how does it work, and how is it connected to True Cost Accounting?

"EOSTA", a prominent Dutch distributor of organic fruits and vegetables, employs TCA to expose the hidden environmental and social costs of food production. Their initiative aims to provide full transparency regarding the real impact of agricultural practices and to promote sustainable consumption through informed choices (EOSTA 2017a).

EOSTA is a trailblazer in the organic food sector, recognized for its innovative approach to sustainability (Skoppek and Karstens 2005). Serving a global market, EOSTA leverages its "Nature & More" transparency initiative to educate consumers about the ecological and social footprint of their products (Skoppek and Karstens 2005). This program emphasizes the importance of understanding the full costs associated with food production, including the often-overlooked environmental and social impacts (Drost and Bandel 2016). By sharing detailed sustainability information, EOSTA influences consumer behavior worldwide, encouraging more responsible and sustainable consumption patterns.

EOSTA's TCA methodology involves a comprehensive assessment of environmental and social costs linked to their products. Key components of their evaluation include (Geoffroy and van Tulder 2020):

- **Carbon footprint**: Calculating the total greenhouse gas emissions produced during the cultivation, processing, and transportation of their organic fruits and vegetables.
- **Water usage**: Measuring the amount of water consumed throughout the production process, particularly in water-scarce regions, and promoting practices that minimize water use.
- **Soil health**: Evaluating the impact of farming practices on soil fertility and biodiversity, emphasizing organic farming methods that enhance soil quality.
- **Labor conditions**: Assessing the social costs by examining the working conditions and fair wages provided to farm laborers, ensuring ethical labor practices.

These were presented in the form of an original image campaign by EOSTA. Greenhouse gases are depicted in the form of a ball, the required soil in the form of a whole heap, the water consumption in a pool or the health costs with a hospital bed directly in the supermarket (Drost and Bandel 2016; cf. Figure 5.1a–d).

EOSTA integrates these findings into their pricing strategies, effectively internalizing external costs to reflect the true societal and environmental value of their products (Geoffroy and van Tulder 2020). The "Nature & More" initiative uses a unique labeling system, where each product is tagged with a QR

Figure 5.1 a–d Graphic information campaign on TCA. Savings through more sustainable production: Soil, CO_2, water, and health (EOSTA et al. 2017) (*Continued*)

Figure 5.1 (Continued)

code. Scanning the QR code allows consumers to access detailed information about the product's environmental and social footprint (Bandel 2007). For instance, a consumer purchasing organic apples can learn about the specific farm where the apples were grown, the farming practices employed, and the overall ecological impact of the product (Pétursson 2013).

EOSTA's implementation of TCA aligns perfectly with a broader understanding of sustainability practices that internalize external costs. By making these hidden costs visible, EOSTA educates consumers and catalyzes market shifts towards more sustainable choices. Their approach demonstrates the feasibility and effectiveness of integrating TCA in the food supply chain, providing valuable insights and a replicable model for other companies (Drost and Bandel 2016).

EOSTA's initiative shows that businesses can take a proactive role in sustainability, going beyond mere compliance to actively promote and facilitate sustainable consumption. The transparency offered by the "Nature & More" labels empowers consumers to make informed decisions, fostering a deeper connection between consumption and its environmental impact. This case study underscores the transformative potential of TCA in driving significant changes in consumer behavior and industry practices globally, highlighting the crucial role of businesses in achieving sustainability goals.

In 2017, the former Prince Charles commended EOSTA for their pioneering work in TCA, particularly highlighting the study "True Cost Accounting in Farming and Finance" (EOSTA 2017b). At the "Harmony in Food and Farming Conference" in Wales, EOSTA's CEO Volkert Engelsman presented the findings, which align with the principles of the "Accounting for Sustainability" project founded by Prince Charles. The Prince, a long-time advocate of organic farming and sustainability, emphasized the critical need to account for the true costs of food production, including impacts on climate, water quality, soil erosion, and human health. He noted that the production of organic apples, for instance, shows a true cost (EOSTA 2017b) decrease of €0.19 per kilogram compared to conventional apples, due to the health benefits and reduced environmental impact. This endorsement by Prince Charles highlights the importance and influence of EOSTA's TCA approach on a global political scale.

5.8 PENNY: True costs in German food retail

Key issue: What was the background of the True Cost Accounting campaign by PENNY, which effects had it, and how does this relate to the broader context of True Cost Accounting?

"PENNY", a major German discount supermarket chain, which is part of the "REWE Group", has implemented a pioneering TCA campaign to disclose the true environmental costs of their products. Conducted in partnership with the University of Greifswald and the Technical Institute of Nuremberg, which have been researching the external costs of agriculture for years under the BMBF research project HoMaBiLe and the EU project FOODCoST. The initiative aims to raise consumer awareness about the hidden ecological impacts of their food choices and promote more sustainable consumption patterns (Universität Greifswald 2023).

PENNY's TCA initiative is notable for its scale and impact, making it a significant case study for national and international audiences interested in sustainable retail practices. As a leading food retailer, PENNY serves millions of consumers across Germany (BVE 2023). Their initiative demonstrates how large-scale retailers can leverage TCA to influence consumer behavior and promote sustainability. By integrating TCA into their pricing strategy, PENNY provides a possible model for other retailers globally, showing how transparency regarding environmental costs can drive more sustainable consumption and lead to necessary societal and political debates regarding the status quo of agri-food systems.

As part of an initial campaign in 2020, PENNY had 18 foodstuffs (nine products, each conventional and organic) scientifically analyzed. The four categories of greenhouse gas emissions, energy, land use change, and reactive nitrogen were mapped. These scientific calculations by the University of Greifswald were then presented in the form of a second price tag (see Figure 5.2) for the environmental impact costs in a store in Berlin as a pure information campaign (Michalke et al. 2020).

Even this small information campaign was associated with a high media response and brought the topic of TCA into the social debate. The information campaign was analyzed in the paper Michalke et al. (2022) and provided the following findings, among others: Through consumer surveys and expert interviews, the research highlights that consumers are interested in true food pricing and are willing to pay these prices to some extent. However, concerns by consumers (survey) and stakeholders (expert interviews) about transparency and social justice, including the fair distribution of wealth and effective communication, pose challenges. Implementing TCA requires socially sensitive measures and supportive legal frameworks to ensure fairness and responsibility among policymakers, practitioners, and consumers (Michalke et al. 2022).

Therefore, based on initial research findings from the information campaign, the next step was taken in the form of an experiment in the implementation of TCA: PENNY's nationwide TCA campaign in 2023, where customers had to pay the true prices, involved a detailed analysis of the environmental

Figure 5.2 Second price tag for true cost of minced meat. True Cost informational campaign at PENNY "Grüner Weg" in Berlin Spandau (Michalke et al. 2022)

costs associated with various products (Stein et al. 2024). The scientific methodology used for calculating the true costs is divided into two main steps:

Step 1: Quantification of emissions, pollutants, and resource use

The first step involves collecting data on the amounts of pollutants and emissions released, as well as resource consumption. This is broken down into several impact categories through life cycle assessments (LCAs), also known as environmental impact assessments, for different food products. These assessments cover all processes involved up to the processing of products. For example, in the case of a vegan schnitzel, the process includes converting wheat protein into a schnitzel. The LCA methodology helps categorize various emissions, pollutants, and resources into 18 impact categories (Stein et al. 2024), which are grouped into four main areas:

- **Health**: Ozone depletion, photochemical oxidant formation, ionizing radiation, human toxicity, particulate matter formation. The health damage is assessed based on the release of harmful substances during production rather than the consumption of unhealthy foods.
- **Water**: Marine ecotoxicity, freshwater eutrophication, freshwater ecotoxicity, marine eutrophication.
- **Soil**: Urban land use, agricultural land use, terrestrial ecotoxicity, terrestrial acidification.
- **Climate**: Greenhouse gas emissions in CO_2e (carbon dioxide, methane, and nitrous oxide) (Wahre-Preise.com 2023).

This step distinguishes between organic and conventional foods, recognizing that organic farming prohibits the use of synthetic fertilizers and pesticides, which positively impacts ecological outcomes (Stein et al. 2024).

Step 2: Monetizing emissions, pollutants, and resource use

The second step involves translating these quantified impacts into monetary values. This is done using damage cost factors that describe the economic consequences of the impacts within the four damage categories: water, soil, climate, and health. These cost factors are often provided by organizations such as the Federal Environment Agency of Germany.

It is important to note that while this methodology approximates the "true price" of food products, it does not fully capture all hidden costs. Both the LCAs (Step 1) and the cost calculations (Step 2) come with uncertainties and assumptions. Not all hidden costs are known, so additional parameters could influence price setting. Thus, TCA offers an approximation of the true prices, highlighting areas where negative effects should be avoided or compensated (Stein et al. 2024).

Before the Initiative in 2023, PENNY initially introduced TCA prices for a select range of products in 2020. These displayed both the actual retail price and the "true cost" price, which included environmental factors.

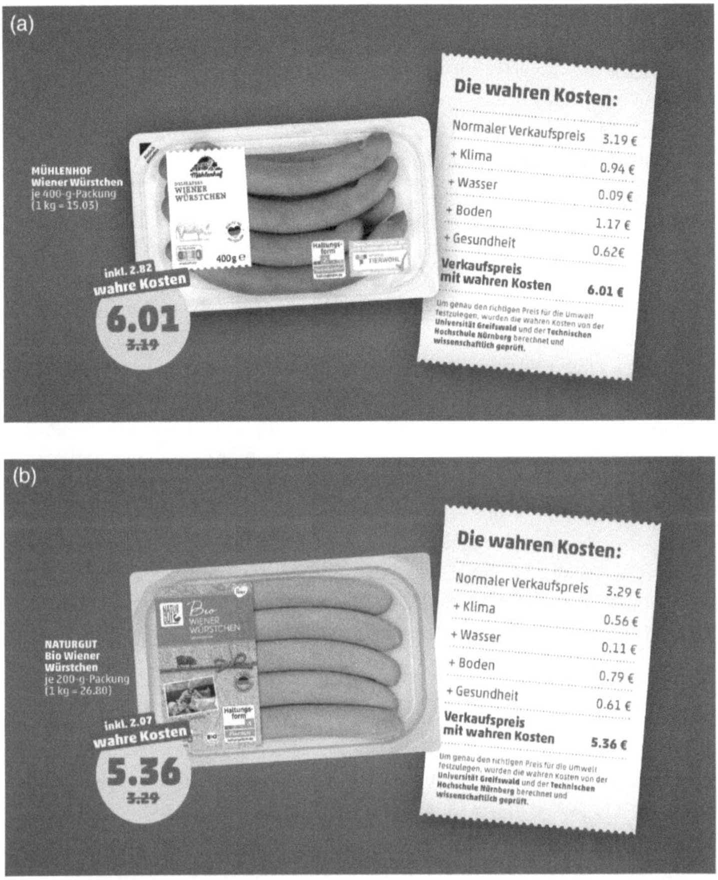

Figure 5.3 a, b TCA information on the True Cost Campaign in all German PENNY stores in 2023. Conventional and organic sausages including their environmental costs (PENNY 2023)

This dual pricing system aimed to inform consumers about the hidden costs of food production and encourage more environmentally friendly choices (Figure 5.3a, b).

In 2023, PENNY expanded this initiative nationwide, incorporating academic research findings to refine their methodology and enhance consumer engagement. This expanded campaign involved detailed educational efforts within store (Ohm 2024), including informational displays and interactive digital content to explain the concept of TCA and its importance. The aim was to deepen consumer understanding and foster a culture of sustainability.

For example, the campaign highlighted the true cost pricing for conventional versus organic apples. The true cost price for conventional apples accounted for the higher environmental impact of pesticide use, higher carbon emissions, and greater water consumption. In contrast, the true cost price for organic apples was lower, reflecting their reduced environmental footprint due to sustainable farming practices (Stein et al. 2024).

The first preliminary results of the scientific evaluation of the campaign were published at the Green Week in Berlin at the beginning of 2024 (which will be published in detail in the paper Semken et al. 2024 in the near future).

Sales impact: During the campaign week from July 31 to August 5, 2023, sales of the featured products generally decreased compared to reference periods. However, the decline was not as steep as expected given the significant price increases, possibly due to effective science communication and the association of additional revenues with donations to a sustainability project. Organic products saw a smaller drop in sales than conventional ones. For instance, the sales of organic mozzarella (49% price increase) fell by 29%, while the sales of conventional mozzarella (74% price increase) (Stein et al. 2024) declined by 43%. Notably, sales of plant-based products increased despite a price hike, likely due to a smaller price increase of 5%.

Donations: A total of €375,000 was donated to the "Zukunftsbauer" project. This sum includes the difference between the sales price and the cost price of the goods, along with an additional €50,000 donation from PENNY (Presseportal 2024).

Customer perception: The campaign generated significant public and media interest, with over 1200 articles published and extensive coverage in TV, radio, and podcasts. A survey with 2255 participants revealed that 64% were aware of the campaign, and it raised awareness about true costs (53%) and sparked discussions on political reforms (38%). This survey on TCA before and after PENNY's campaign also revealed the following when comparing approval of the measures: Support for implementing TCA measures, such as displaying true cost information or adding a surcharge, has decreased compared to the pre-campaign survey (Stein et al. 2024).

Survey results: Post-campaign surveys showed that consumers recognized higher true costs of products. However, the survey participants did not differentiate significantly between different product groups in their assessments of the high level of externalities. The campaign highlighted the public's realization that true costs are generally higher than initially perceived.

Regional sales differences: Sales of the True Cost products declined nationwide, with more significant drops in the new federal states (up to 70%) and smaller decreases in western (North Rhine-Westphalia) and southern (border regions with Austria and Switzerland) Germany. These regional variations could be attributed to factors such as income levels, interest in sustainability, or previous demand.

PENNY's TCA project aligns seamlessly with the principles of TCA by internalizing environmental costs and making them visible to consumers. This approach not only raises awareness but also incentivizes more sustainable purchasing decisions by highlighting the real cost of food production. The initiative's success demonstrates the effectiveness of TCA in the retail sector, offering a compelling case for its broader adoption (Ohm 2024).

By taking these steps, PENNY contributes significantly to environmental sustainability while setting a precedent for other retailers worldwide. Their comprehensive approach combines a detailed cost assessments with consumer education and provides a robust model for integrating TCA into retail practices. This case study exemplifies the potential of TCA to drive meaningful change in consumer behavior and industry standards, reinforcing the critical role of businesses in advancing the sustainability agenda (Wahre-Preise.com 2023).

These detailed case studies illustrate the transformative power of TCA in promoting sustainable business practices and consumer awareness, contributing significantly to the global sustainability goals.

However, it is of great importance not to leave it at this campaign (which was also seen by some as a greenwashing campaign by PENNY), but to integrate externalities, for example, into the accounting of companies and the associated corporate strategy. It remains to be seen whether PENNY and other food retailers will take this step proactively or only after the appropriate legal framework has been created.

References

Abel, G. (2020). Der kritische und mündige Verbraucher. In: Hellmann, K. U., Klein, A., & Baule, B. (eds) *Verbraucherpolitik von unten* (p. 275–292). Springer Nature.

Abel, G., & Bätzing, M. (2020). Einfach klimagerechter leben. *Forschungsjournal soziale Bewegungen, 33*(1), 112–124. https://doi.org/10.1515/fjsb-2020-0010

Atmosfair. (2021). *Annual report 2021.* Available at: https://www.atmosfair.de/wp-content/uploads/atmosfair-2021-annual-report-en-1.pdf (Accessed July 18, 2024)

Atmosfair. (2022). *Jahresberichte.* Available at: https://www.atmosfair.de/wp-content/uploads/atmosfair-2022-jahresbericht-web.pdf (Accessed July 18, 2024)

Atmosfair. (2023). *Emission calculation: Calculating the carbon emissions of flights, considering factors such as aircraft type, flight distance, and load factors.* Available at: https://www.atmosfair.de/wp-content/uploads/flight-emissionscalculator-documentation-calculationmethodology.pdf (Accessed July 18, 2024)

Atmosfair. (2024a). *Overview over climate protection projects.* Available at: https://www.atmosfair.de/de/klimaschutzprojekte/ (Accessed July 18, 2024)

Atmosfair. (2024b). *Waldschutzprojekte.* Available at: https://www.atmosfair.de/de/standards/waldschutzprojekte/ (Accessed July 18, 2024)

Bandel, T. (2007). *Establishing a comprehensive, multi-level linkage between small farmers in Egypt and consumers in export markets. 3rd International Conference on Linking Markets and Farmers: Exploring leading practices to promote economic growth in rural India.*

Bilharz, M. (2005). *Nachhaltigkeits-Marketing in Theorie und Praxis* (1st ed., 181–196). Deutscher Universitäts-Verlag.

Bioland. (2024a). *About Bioland.* Available at: https://www.bioland.de/about-bioland (Accessed July 18, 2024)

Bioland. (2024b). *Amendments to the standard of March 18/19, 2024.* Available at: https://www.bioland.de/fileadmin/user_upload/Verband/Dokumente/Richtlinien_fuer_Erzeuger_und_Hersteller/Bioland-Richtlinien_englisch_2024-03_.pdf (Accessed July 18, 2024)

Braatz, N. (2022). *Klimaneutral-Label: Recherche deckt fragwürdige Methoden auf.* Utopia. Available at: https://utopia.de/news/klimaneutral-siegel-anbieter-fragwuerdige-methoden/ (Accessed July 18, 2024)

Bundesvereinigung der Deutschen Ernährungsindustrie e.V. (BVE). (2023). *Jahresbericht Ernährungsindustrie 2023.* Available at: https://www.bve-online.de/download/bve-jahresbericht-ernaehrungsindustrie-2023 (Accessed June 18, 2024)

BMU. (2017). *Carbon pricing: Using market-based mechanisms to mitigate climate change.* Federal Ministry for the Environment, Nature Conservation and Nuclear Safety. Available at: https://www.BMU.de/fileadmin/Daten_BMU/Pools/Broschueren/treibhausgasemissionen_en_bf.pdf (Accessed July 18, 2024)

ClimatePartner. (2024a). *Protocol certification.* Available at: https://ftp.climatepartner.com/download/CP_Protocol_certification_EN.pdf (Accessed June 18, 2024)

ClimatePartner. (2024b). *Projekt-Technologien der Klimaschutzprojekte.* Available at: https://www.climatepartner.com/de/klimaschutzprojekte/projekt-technologien (Accessed July 18, 2024)

ClimatePartner. (2024c). *Quality guidelines for carbon neutral companies, products, and services.* Available at: https://www.climatepartner.com/sites/default/files/documents/WEB_CP_Protocol_carbon_neutrality_EN.pdf (Accessed June 18, 2024)

Drost, R., & Bandel, T. (2016). *True Cost Accounting for food, farming & finance.* Social.eu. Available at: https://www.socila.eu/wp-content/uploads/sites/5/2018/01/True-Cost-Accounting-for-Food-Farming-and-Finance.pdf (Accessed June 18, 2024)

Du bist hier der Chef. (2021). *Unsere Milch. Die Ergebnisse im Detail.* Available at: https://dubisthierderchef.de/produkte/frischmilch (Accessed June 18, 2024)

Du bist hier der Chef. (2024). *Die Geschichte unserer Verbraucher-Initiative.* Available at: https://dubisthierderchef.de/geschichte-der-initiative (Accessed June 18, 2024)

EOSTA. (2017a). *True Cost Accounting-Pilotstudie legt wahre Kosten der Lebensmittelproduktion für Mensch und Umwelt offen.* Available at: https://www.eosta.com/de/aktuell/true-cost-accounting-pilotstudie-legt-wahre-kosten-der-lebensmittelproduktion-fuer-mensch (Accessed June 18, 2024)

EOSTA. (2017b). *Prinz Charles lobt Studie "True Cost Accounting in farming and finance".* Available at: https://www.eosta.com/de/aktuell/prinz-charles-lobt-studie-true-cost-accounting-in-farming-and-finance (Accessed June 18, 2024)

EOSTA, Soil & More, EY, Triodos Bank, & Hivos. (2017). *True Cost Accounting for Food, Farming and Finance (TCA-FFF).* Available at: https://www.natureandmore.com/sites/www.natureandmore.com/files/teler/eosta_report_-_true_cost_accounting_for_food_farming_and_finance.pdf

European Commission (EUCOM). (2024a). *Bio-Logo.* Available at: https://agriculture.ec.europa.eu/farming/organic-farming/organic-logo_de (Accessed July 18, 2024)

European Commission (EUCOM). (2024b). *Trade in organics.* Available at: https://agriculture.ec.europa.eu/farming/organic-farming/trade_en (Accessed July 18, 2024)

European Network for Rural Development. (2022). *Bioland: Organic farming case study*. European Commission. Available at: https://ec.europa.eu/enrd/sites/default/files/enrd_publications/case_study_bioland.pdf (Accessed July 18, 2024)

European Union (EU). (2018a). Regulation (EU) 2018/848 of the European Parliament and of the council of 30 May 2018 on organic production and labelling of organic products and repealing council regulation (EC) no 834/2007. *Official Journal of the European Union*, *150*, 1–92.

European Union. (2018b). Regulation (EU) 2018/848 of the European Parliament and of the council of 30 May 2018 on organic production and labelling of organic products and repealing council regulation (EC) no 834/2007. *Official Journal of the European Union*, *150*, 1–92.

European Union. (2018c). Regulation (EU) 2018/848 of the European Parliament and of the council of 30 May 2018 on organic production and labelling of organic products and repealing council regulation (EC) no 834/2007. *Official Journal of the European Union*, *150*, 1–92.

Fischer, T., & Knuth, H. (2023). *CO2-Zertifikate: Grün getarnt. DIE ZEIT*. Available at: https://www.zeit.de/2023/04/co2-zertifikate-betrug-emissionshandel-klimaschutz (Accessed June 18, 2024)

Foodwatch. (2022). *DER GROSSE KLIMA-FAKE*. Available at: https://www.foodwatch.org/fileadmin/-DE/Themen/Klimaluegen/Report_Klima_Claims/Klima_Report_2022_2.pdf (Accessed July 18, 2024)

Geoffroy, P., & van Tulder, R. (2020). *Sustainable Eosta: Leading Organic Trade by Valuing People and Environment*. RSM Case Development Centre. Available at: http://hdl.handle.net/1765/130661 (Accessed August 1, 2024)

Gojdka, V., & Timmler, V. (2019). *"Fliegen und das gute Gewissen: Wie sinnvoll ist die CO2-Kompensation?"*. Süddeutsche Zeitung. Available at: https://www.sueddeutsche.de/wirtschaft/atmosfair-flug-kompensieren-1.4459405 (Accessed June 19, 2024)

Grohmann, P., Dietze, V., Krämer, C., & Feindt, P. (2023). Governance and trust in sustainability-based agri-food value chains. A comparative analysis of five cases in Germany. *Frontiers in Sustainable Food Systems*, *7*, 1–14. https://doi.org/10.3389/fsufs.2023.1130895

Lebensmittelklarheit. (2022). *EU-Logo und Herkunftsangaben für Bio-Lebensmittel*. Available at: https://www.lebensmittelklarheit.de/informationen/eu-logo-und-herkunftsangaben-fuer-bio-lebensmittel (Accessed July 18, 2024)

Lefebvre, D., Williams, A. G., Kirk, G. J. D., Paul, J. B., Meersmans, J., Silman, M. R., Román-Danobeytia, F., Farfan, J., & Smith, P. (2021). Assessing the carbon capture potential of a reforestation project. *Scientific Reports*, *11*, 19907.

Leopold, S. (2020). *"Du bist hier der Chef" vermarktet Milch nach Verbraucherwünschen*. Agrarheute. Available at: https://www.agrarheute.com/land-leben/chef-vermarktet-milch-verbraucherwuenschen-567703 (Accessed June 18, 2024)

Michalke, A., Gaugler, T., & Stoll-Kleemann, S. (2020). *Abschlussbericht zum Forschungsprojekt How much is the dish? - True Cost Accounting von Umweltfolgekosten und "wahre Preisschilder" in Deutschland*. Available at: https://geo.uni-greifswald.de/storages/uni-greifswald/fakultaet/mnf/geowissenschaften/Arbeitsbereiche_Geographie/Nachhaltigkeitswissenschaften/05_Projekte/HoMaBiLe/Endbericht_PENNY_WahrePreise.pdf

Michalke, A., Köhler, S., Messmann, L., Thorenz, A., Tuma, A., & Gaugler, T. (2023). True cost accounting of organic and conventional food production. *Journal*

of Cleaner Production, *408*, 137134. https://www.sciencedirect.com/science/article/pii/S0959652623012921

Michalke, A., Stein, L., Fichtner, R., Gaugler, T., & Stoll-Kleemann, S. (2022). True cost accounting in agri-food networks: A German case study on informational campaigning and responsible implementation. *Nachhaltigkeitswissenschaft*, *17*(6), 2269–2285.

NDR. (2022). *Bio-Siegel: Welche Standards gelten für Produkte?*. Available at: https://www.ndr.de/ratgeber/verbraucher/Bio-Siegel-Welche-Standards-gelten-fuer-Produkte,biosiegel104.html (Accessed June 18, 2024)

NewClimate Institute. (2024). *Corporate climate responsibility monitor 2024*. Available at: https://newclimate.org/sites/default/files/2024-04/NewClimate_CCRM2024.pdf (Accessed June 18, 2024)

Oebel, B., Stein, L., Michalke, A., Stoll-Kleemann, S., Gaugler, T. (2024). Towards true prices in food retailing: The value added tax as an instrument transforming agri-food systems. *Sustainability Science*.

Ohm. (2024). *Wunsch nach Transparenz von Lebensmitteln ist messbar gewachsen – Ergebnisse der Wahre-Kosten-Kampagne mit PENNY liegen vor*. Pressemitteilung. Available at: https://www.th-nuernberg.de/pressemitteilung/5207-wunsch-nach-transparenz-von-lebensmitteln-ist-messbar-gewachsen-ergebnisse-der-wahre-kosten-kampagne-mit-penny-liegen-vor/ (Accessed June 18, 2024)

PENNY. (2023). *Wahre Kosten*. Available at: https://www.penny.de/aktionen/wahrekosten

Pétursson, J. (2013). EDUARDO'S APPLES. *Ethnologia Europaea*, *43*(2), 17–29.

Presseportal. (2024). *Uni Greifswald und TH Nürnberg legen Auswertung von PENNY-Projekt "Wahre Kosten 2023" vor Ziele erreicht - Diskussionsanstoß gegeben - Fakten statt gesellschaftlich erwünschten Antworten*. Available at: https://www.presseportal.de/pm/59659/5698028 (Accessed June 19, 2024)

Probst, B., Toetzke, M., Kontoleon, A., Diaz Anadon, L., & Hoffmann, V. H. (2023). *Systematic review of the actual emissions reductions of carbon offset projects across all major sectors*.

Sacchi, R., Becattini, V., Gabrielli, P., Cox, B., Dirnaichner, A., Bauer, C., & Mazzotti, M. (2023). How to make climate-neutral aviation fly. Nat Commun 14, 3989.

Semken, C., Michalke, A., Stein, L., van Sambek, F., Varela Seone, S., Oebel, B., Gaugler, T., & Allcott, H. (forthcoming - under review). *Optimal green retailing: Theory and evidence*. National Bureau of Economic Research.

Skoppek, H., & Karstens, B. (2005). Nachhaltigkeits-Marketing Eines Europäischen Großhandelsunternehmens Am Beispiel Von EOSTA Und „Nature & More". In: Belz, F. M., & Bilharz, M. (eds) *Nachhaltigkeits-Marketing in Theorie Und Praxis* (1st ed., 181–196). Deutscher Universitäts-Verlag.

Spiegel Online. (2020). *Atmosfair, Myclimate, Primaklima, Klima-Kollekte: Wie kompensiere ich meine Reise am besten?* Der Spiegel. Available at: https://www.spiegel.de/reise/atmosfair-myclimate-primaklima-klima-kollekte-wie-kompensiere-ich-meine-reise-am-besten-a-b3fa79ed-49d6-4c98-bec5-c64d3742d208 (Accessed July 18, 2024)

Stein, L., Michalke, A., Gaugler, T., & Stoll-Kleemann, S. (2024). Sustainability science communication: Case study of a true cost campaign in Germany. *Sustainability*, *16*(9), 3842.

Umweltbundesamt. (2020). *Guide to voluntary CO2 offsetting through climate protection projects*. Available at: https://www.umweltbundesamt.de/sites/default/files/

medien/376/publikationen/guide_voluntary_co2_offsetting_through_climate_protection_projects_bf_2020_09_14.pdf (Accessed July 18, 2024)

Umweltbundesamt. (2021). *Umweltschonender Luftverkehr*. Available at: https://www.umweltbundesamt.de/publikationen/umweltschonender-luftverkehr (Accessed June 19, 2024)

Universität Greifswald. (2023). *Echte Kostenrechnung für Lebensmittel bei Discountern*. Available at: https://www.uni-greifswald.de/en/university/information/current-news/details/n/true-cost-accounting-for-food-at-discount-supermarkets-new64d5eed44a764305402808/ (Accessed June 18, 2024)

Wahre-Preise.com. (2023). *Wahre Preise*. Available at: http://wahre-preise.com/ (Accessed June 19, 2024)

We Care. (2023). *We Care Social, sustainable and fair management in food companies Certification Standard, Version 1.3*. p 41 ff. Available at: https://we-care-siegel.org/companies/?lang=en (Accessed July 18, 2024)

We Care. (2024). *We Care – das Siegel für nachhaltiges Unternehmens- und Lieferkettenmanagement*. Available at: https://we-care-siegel.org/we-care-siegel/ (Accessed July 18, 2024)

West, T. A. P., Wunder, S., Sills, E. O., Börner, J., Rifai, S. W., Neidermeier, A. N., Frey, G., & Kontoleon, A. (2023). Action needed to make carbon offsets from forest conservation work for climate change mitigation. *Science*.

Part IV

Conclusion

6 Implications of TCA for Different Stakeholders

Abstract
In the previous chapters, we have drawn a comprehensive picture of True Cost Accounting (TCA) as a valuable possibility to enable companies adopting a more sustainable business model. This final chapter carves out the stakeholders' roles in the context of TCA and summarizes the main advantages and challenges of TCA. As already indicated in the previous chapters, unlike traditional cost accounting, which is typically internally focused and mostly only for internal decision-makers, TCA has close links to external reporting and is meaningful for different stakeholders. Therefore, its successful implementation also depends on involving these stakeholders. To keep the discussion parsimonious, we focus on companies, customers, suppliers, regulators, investors, and educational institutions.

Key takeaways

After reading this chapter, you will:

- Understand the main stakeholders' relevance and roles in implementing and applying True Cost Accounting (TCA) successfully
- Have an overview of the advantages and challenges when assessing the reasonability of implementing TCA

6.1 Role of key stakeholders

Key issue: Which role do the main stakeholders play to implement and apply TCA successfully?

Companies have a central position in the implementation of True Cost Accounting (TCA), not only because they are ultimately the entities that have to implement this accounting system. Also, a critical mass of companies is needed to enable a comprehensive (and maybe even compulsory) implementation and to use the information generated by TCA in the long term. Only then can TCA have an effect on economic activity on a larger scale. To achieve this

DOI: 10.4324/9781003345190-10

goal, companies need an incentive to expand their cost accounting systems accordingly and implement a more comprehensive resource consumption assessment approach. An important impetus here must come from regulation, which will be addressed separately below. However, TCA also offers advantages for companies. Based on Bebbington et al. (2001), Antheaume (2004) already discusses four learnings, which can be derived from research on experiments with true/full cost accounting: First, the process of TCA as such results in a learning experience unveiling potential for process improvement from which the company can benefit. At the time, when these learning experiences were observed, the discussion about sustainable business was not as advanced as it is today, and there was much less regulatory intervention in the activities of companies aimed at sustainable corporate governance. It can therefore be assumed that the learning effects observed at that time can be even more valuable for companies in the current regulatory and economic situation. This includes the identification of process inefficiencies and unnecessary resource consumption. Second, these learning processes additionally indicate the limits of improvements, which can lead to the conclusion that certain aims of sustainable development cannot be reached, at least in the short- to medium-term. This insight might be disappointing at first, but it helps to direct a company's limited resources in an early stage toward those goals that are reasonably attainable. Third, the calculated external costs often amount to a considerable sum, which actually can "turn a profit to a loss" (Antheaume 2004, p. 445). Focusing on environmental cost accounting, Christ and Burritt (2013) argue that organizations are often surprised when they realize how significant environmental costs are. This learning helps to sensitize the entire organization to the fact that the evaluation of externalities is not just a marginal phenomenon that deals with marginal resource consumption, but that it reveals relatively extensive, but hitherto unconsidered, consumptions of natural resources. This, in turn, creates an awareness that efficient use of resources is important at all points in the production chain and that everyone in the organization can make an impactful contribution. Finally, Antheaume (2004) mentions that standards are lacking for the applied methods, making it impossible to compare companies. Unfortunately, this observation still holds true 20 years later. It is therefore not only up to regulation but also up to the companies themselves to agree on certain standards, e.g., coordinated across the industry representatives.

Apart from these four learnings, other scholars additionally point to a possible competitive advantage, which can be generated through the implementation and maintenance of some kind of environmental management accounting (e.g., Bouma and van der Veen 2002). As indicated by Jasinski et al. (2015, p. 1127) based on Zokaei et al. (2013): "Toyota saved approximately 38 billion yen (equivalent to half a billion US dollars) between 2008 and 2010 mainly by reducing energy consumption, reducing waste processing costs, selling recyclable goods and utilizing other environmentally-friendly technologies".

Of course, companies are still better off if they sell their products at a low price, and the neglect of externalities certainly supports this. However, on the other hand, there seems to be a growing number of companies for which a sustainable business model is significant. For these companies, TCA offers the possibility to gain information that can be used for external and internal communication purposes and that constitutes an important component of a web of mechanisms, which guide corporate activities in the intended direction. Or as Riccaboni and Leone (2010, p. 131) put it: "In other words, social reporting is not enough since environmental and social profiles should be incorporated within planning process, policy decisions, capital allocation and performance evaluation". TCA can be such an incorporating mechanism. Furthermore, it is capable of identifying possible trade-offs between objectives when evaluating investment alternatives (Jasinski et al. 2015), i.e., it is compatible with all three dimensions of the triple bottom line. This compatibility comes at a cost, i.e., there is not the one and only calculation method, resulting in a range of values assigned to externalities for which no market values exist. Yet, "as Rubenstein (1992) concluded, it is still better to be approximately right than precisely wrong" (Jasinski et al. 2015, p. 1136).

Customers are another important stakeholder group in the context of TCA, which can be further differentiated in business and end customers. Business customers are other companies that are subject to the same market mechanisms and regulatory requirements as the focal firm. Therefore, for them, the same argumentation holds as previously discussed. If they are interested in truly achieving a sustainable value creation, they are interested in cooperating with sustainable suppliers, and TCA can help them to make the supply chain more sustainable. However, end customers follow another economic logic and therefore have to be discussed separately. To derive the relevance of TCA for different end customers, we differentiate them into four groups (see Figure 6.1). This differentiation is based on two dimensions. The first dimension comprises the financial solvency. Broadly speaking, end customers can dispose over a high or a low financial solvency in the sense that they are capable of paying higher prices resulting from the incorporation of social and ecological externalities. The second dimension covers the importance that end customers assign to an ecological and socially fair production of those goods that they consume. Also here, we differentiate broadly between a high relevance assigned versus a low relevance assigned. These two dimensions are important because they indicate that individuals might be willing to pay higher prices for externalities but are not capable to do so due to their economical situation, while others could afford to pay for this but are unwilling to, as they are not interested in ecological and social issues. The application of TCA and dissemination of information derived from it can exert an impact on all four groups. Members of the first group, i.e., people who are financially solvent and dedicated to sustainability, need information helping them to make informed decisions to achieve their goal of a truly sustainable lifestyle. These

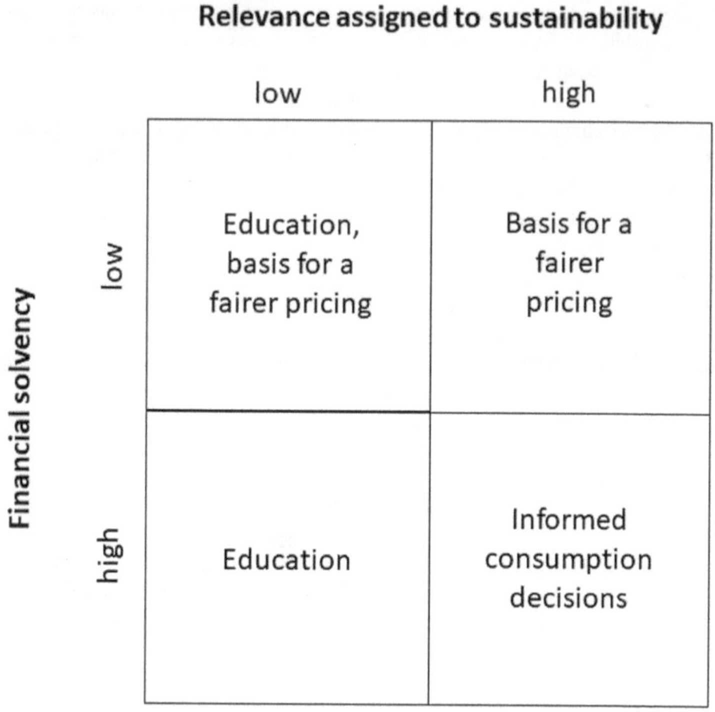

Figure 6.1 Relevance assigned to TCA by different customers

people are not deterred by possibly high prices due to the internalization of externalities as they can afford to pay them, but by attempts of greenwashing and unsustainable production processes. Thus, if a company depends to a high degree on these customers, TCA can help to provide the information that these customers highly value. However, members of the second group, i.e., people who place a high value on sustainability but who cannot afford to buy high-priced products, are also interested in the information provided by TCA, because this information not only indicates the true costs of one single product, but it also allows for the comparison between different products or production processes. For example, the analysis by Pieper et al. (2020) indicates major differences between plant-based versus meat-based food and organic versus conventional food production. This information can be the starting point for a discussion around fair pricing of food or other products, resulting, e.g., in changes of national or international grant practices, which, in turn, lead to a lowering of prices of organic versus conventional food. Thus, in the long run, these end customers also can profit from the outcomes of TCA, as it can help to resolve some of the inequalities while making them measurable by drawing broader societal attention to them. In contrast, individuals who are financially solvent but not interested in sustainability in the context of their consumption might be no relevant audience for TCA information at first glance. Yet, as indicated by the results of a study conducted by Ignat and Chankov (2020) on the effect of sustainability information in the context of e-commerce the confrontation with information comparing different production practices and products (with similar functionalities) might exert an educational effect on them, i.e., they might get a better understanding about why sustainability is a relevant dimension within business and market contexts. As a consequence, some of them might switch over to those end customers who assign certain value to sustainable production practices and are willing to pay for it. The final group, those people who are uninterested in sustainability and who cannot afford to pay higher prices, might benefit in two ways. First, they can learn more about the relevance and the impact of sustainability in the context of consumption and, second, they also might profit from the outcomes of societal debates induced by the insights gained from TCA. In sum, TCA can make sense for all of these end customers.

Suppliers get more and more in the focus of sustainable management activities partly due to changes in regulation, such as the German Supply Chain Due Diligence Act mentioned in Chapter 4. At the same time, achieving a sustainable supply chain is a challenging endeavor, as particularly at the origin of these supply chains processes and practices are getting less and less transparent. This fact also limits the scope of TCA. Thus, if a company wants to implement a meaningful environmental and social information system, it depends on the cooperation and transparency of at least its most important suppliers. Consequently, TCA is most effective if it is implemented not only in the focal company but also when relevant suppliers are willing to engage in it. Depending on the location of these suppliers, their position along the complete supply chain, and their individual interest in sustainability, these suppliers may also have a genuine interest in participating in TCA. However, for those suppliers who are not interested in sustainability, who are not affected by regulation, or who just lack resources to deal with sustainability issues, the involvement in TCA along a supply chain is unreasonable. This will make it difficult for the focal company to get them on board. Consequently, companies intending to implement TCA should screen their suppliers beforehand to identify those that can be reasonably involved.

Regulators and standard setters are the most important stakeholder to provide the prerequisites for extracting the best out of TCA. As discussed in previous chapters, regulation, particularly in the European Union, changes the rules for external reporting to direct investment activities toward green and socially responsible investments. Therefore, the implementation of TCA in companies is also in the European Union's interest. However, so far, little attempts have been made to change cost accounting, as the regulation's focus still lies on external reporting and cost accounting concentrates mainly on internal use. But as Adams and Larrinaga-González (2007, p. 339) indicate based on Adams (2004): “increased sustainability reporting does not necessarily mean improved accountability”. Therefore, if regulators want to bring about real change, they also have to focus on management control activities internal to companies. TCA can help to align these activities with the wished-for sustainable development. However, companies will not adopt TCA in competitive market environments (Burritt 2005), as this might disadvantage them compared to companies that do not care about environmental and social issues, except for when they are legally obligated to. Currently many companies are already heavily burdened with the implementation of sustainability reports. An obligation to implement additional components of sustainable accounting, such as TCA, would certainly lead to excessive demands. In the medium term, however, it should be considered whether mandatory implementation of TCA could make sense, at least for larger, resource-rich companies with a correspondingly high sustainability impact. On a positive note, such an approach can also help to standardize sustainability reporting, as TCA can promote standardization in the measurement of sustainable activities (El-Hage Scialabba and Obst 2021).

Investors are another important stakeholder group that can exert pressure on companies to engage in sustainable development. However, as the experience of the last decades shows, they usually profit most by investing in less sustainable companies, and therefore their interest in paying for externalities by pricing them accordingly is rather low. The changed legal situation of EU countries, reflected in the taxonomy and a series of reporting directives that are being transposed into national law, attempts to take this observation into account. Investors are thus increasingly becoming the focus of regulators in order to achieve the desired sustainable transformation of the overall economy. If they take these changes seriously, they need reliable information to guide their investments toward sustainable companies. TCA can provide this information and should therefore be in the interest of investors to find its way into corporate management.

Educational institutions play a special role in the introduction of innovative management tools. The idea behind TCA is not new, but it has obviously not yet been established in business practice. This is certainly also due to the fact that prospective managers and controllers have hardly been confronted with it in their training to date. For example, the results of a survey by Boulianne et al. (2018) reveal that although the French government requires sustainable reporting, the state-supported training of accountants hardly mentions it. TCA has not yet found its way into regular cost accounting training, and it is also a niche topic in continuing education. If its potential for companies is to be leveraged, more intensive efforts are needed from all types of training and education institutions to include TCA more prominently in their curricula. This also comprises professional accounting associations, which are dedicated to keeping their members up-to-date by providing information on new developments in this area, as highlighted already by Zvezdov and Schaltegger (2013). This does not only include the teaching of the basic techniques of TCA but also the showing of technical solutions and software programs that make implementation as efficient and cost-effective as possible. Already Burritt (2005) sees this as an important point to enable smaller companies in particular to introduce such accounting systems.

6.2 Key aspects related to TCA

Key issue: What are the main aspects to keep in mind when thinking about implementing TCA?

In the previous chapters, we aimed to provide a comprehensive perspective on TCA. This final section is dedicated to highlight the most important advantages but also challenges of TCA.

The first advantage is related to the reasons behind the development of TCA. Classical economic evaluation systems focus on resource consumption, processes, and outcomes that can be monetized rather easily and that result in cash in- and outflows for the economic entities, like companies or consumers. This focus has shaped the thinking of management and regulators for decades. However, it has the drawback of missing impactful natural and social resource consumptions, which lead to major ecological and social problems and eventually also endanger the economic system as such. Therefore, more powerful assessment instruments are needed, which can account for these externalities in a comprehensive and understandable manner. TCA poses one of such tools and *can exert an impact on both corporate internal decision making and on the decision processes of stakeholders*, like customers or suppliers, to foster the sustainable development of companies. As a consequence, it can also provide information needed for reporting purposes induced by regulation changes.

Second, TCA has long been under development, which yielded *a broad range of different calculation approaches*. Therefore, its application requires further streamlining and is resource intensive. However, as different calculation methods are sensible, it also can deal with various industries and particularities in companies. This flexibility makes it a useful, albeit not free of charge, starting point for sustainable management.

Third, TCA can *be combined with various other instruments dedicated to sustainable management*. Therefore, before introducing TCA in the company, it is worth taking stock of the existing control systems in this area in order to identify interfaces and make the use of data as efficient as possible. This approach is time consuming and requires resources but will eventually pose a comprehensive management tool that provides the company with information needed to develop more sustainably and also to get more resilient in turbulent times.

Fourth, TCA is an interdisciplinary approach that requires the *collaboration of several departments*. Therefore, a project team is needed, staffed with members of various departments, particularly the accounting and the sustainability departments, but also experts from other areas. Departments must be willing and enabled to cooperate and financial resources are needed to keep the process going. Moreover, this need for cooperation makes implementation time-consuming, but this also bears the possibility to improve communication between the departments, eliminate obstacles to cooperate more efficiently in general, and combine the knowledge existing in the company in a fruitful manner, which would not have been possible or certainly more difficult without these processes.

In sum, TCA is not an easy, inexpensive endeavor, but it can help to make a company more sustainable and more efficient and thereby provide the ground for an enduring competitiveness.

6.3 Review questions

1 What prerequisites should exist for a company if it wants to implement TCA?
2 Why is regulation important to promote the use of TCA?
3 What types of stakeholders are there, and how is TCA helpful for each?
4 To what extent is the supply chain relevant for the implementation of TCA?
5 To what extent do educational institutions play a special role in introducing TCA to the broader business community?
6 What do you think is the biggest advantage of TCA?
7 What do you think is the biggest disadvantage of TCA?

References

Adams, C. A. (2004). The ethical, social and environmental reporting-performance portrayal gap. *Accounting, Auditing & Accountability Journal, 17*(5), 731–57.

Adams, C. A., & Larrinaga-González, C. (2007). Engaging with organisations in pursuit of improved sustainability accounting and performance. *Accounting, Auditing & Accountability Journal, 20*(3), 333–355.

Antheaume, N. (2004). Valuing external costs – From theory to practice: Implications for full cost environmental accounting. *European Accounting Review, 13*(3), 443–464.

Bebbington, J., Gray, R., Hibbit, C., & Kirk, E. (2001). *Full Cost Accounting: An Agenda for Action*. Association of Chartered Certified Accountants.

Boulianne, E., Keddie, L. S., & Postaire, M. (2018). (Non)coverage of sustainability within the French professional accounting education program. *Sustainability Accounting, Management and Policy Journal, 9*(3), 313–335.

Bouma, J. J., & van der Veen, M. (2002). Wanted: A Theory for Environmental Management Accounting. In: Bennett, M., & Bouma, J. J. (eds) *Environmental Management Accounting: Informational and Institutional Developments. Eco-Efficiency in Industry and Science* (Vol. 9, 279–290). Springer.

Burritt, R. L. (2005). Challenges for Environmental Management Accounting. In: Rikhardsson, P. M., Bennett, M., Bouma, J. J., & Schaltegger, S. (eds) *Implementing Environmental Management Accounting: Status and Challenges* (19–44). Springer.

Christ, K. L., & Burritt, R. L. (2013). Environmental management accounting: The significance of contingent variables for adoption. *Journal of Cleaner Production, 41*, 163–173.

El-Hage Scialabba, N., & Obst, C. (2021). From Practice to Policy – New Metrics for the 21st Century. In: Gemmill-Herren, B., Baker, L. E., & Daniels, P. A. (eds) *True Cost Accounting for Food – Balancing the Scale* (13–25). Routledge.

Ignat, B., & Chankov, S. (2020). Do e-commerce customers change their preferred last-mile delivery based on its sustainability impact? *The International Journal of Logistics Management, 31*, 521–548.

Jasinski, D., Meredith, J., & Kirwan, K. (2015). A comprehensive review of full cost accounting methods and their applicability to the automotive industry. *Journal of Cleaner Production, 108*, 1123–1139.

Pieper, M., Michalke, A., & Gaugler, T. (2020). Calculation of external climate costs for food highlights inadequate pricing of animal products. *Nature Communications, 11*, 1–13.

Riccaboni, A., & Leone, E. L. (2010). Implementing strategies through management control systems: The case of sustainability. *International Journal of Productivity and Performance Management, 59*(2), 130–144.

Rubenstein, D. B. (1992). Bridging the gap between green accounting and black ink. *Accounting, Organizations and Society, 17*, 501–508.

Zokaei, K., Lovins, H., Wood, A., & Hines, P. (2013). *Creating a Lean and Green Business System: Techniques for Improving Profits and Sustainability*. CRC Press.

Zvezdov, D., & Schaltegger, S. (2013). Sustainability Accounting. In: Idowu, S.O., Capaldi, N., Zu, L., & Gupta, A.D. (eds) *Encyclopedia of Corporate Social Responsibility* (2363–2370). Springer.

Index

Note: – *Italicized* page references refer to figures.

FSC
www.fsc.org
MIX
Papier | Fördert
gute Waldnutzung
FSC® C083411

Kolibri 360 GmbH
Ferdinand-Jühlke-Straße 7
99095 Erfurt, Deutschland
produktsicherheit@kolibri360.de